Springer Theses

Recognizing Outstanding Ph.D. Research

Aims and Scope

The series "Springer Theses" brings together a selection of the very best Ph.D. theses from around the world and across the physical sciences. Nominated and endorsed by two recognized specialists, each published volume has been selected for its scientific excellence and the high impact of its contents for the pertinent field of research. For greater accessibility to non-specialists, the published versions include an extended introduction, as well as a foreword by the student's supervisor explaining the special relevance of the work for the field. As a whole, the series will provide a valuable resource both for newcomers to the research fields described, and for other scientists seeking detailed background information on special questions. Finally, it provides an accredited documentation of the valuable contributions made by today's younger generation of scientists.

Theses are accepted into the series by invited nomination only and must fulfill all of the following criteria

- They must be written in good English.
- The topic should fall within the confines of Chemistry, Physics, Earth Sciences, Engineering and related interdisciplinary fields such as Materials, Nanoscience, Chemical Engineering, Complex Systems and Biophysics.
- The work reported in the thesis must represent a significant scientific advance.
- If the thesis includes previously published material, permission to reproduce this must be gained from the respective copyright holder.
- They must have been examined and passed during the 12 months prior to nomination.
- Each thesis should include a foreword by the supervisor outlining the significance of its content.
- The theses should have a clearly defined structure including an introduction accessible to scientists not expert in that particular field.

More information about this series at http://www.springer.com/series/8790

Itzik Kapon

Searching for 2D Superconductivity in $La_{2-x}Sr_xCuO_4$ Single Crystals

Doctoral Thesis accepted by
the Technion - Israel Institute of Technology,
Haifa, Israel

 Springer

Author
Dr. Itzik Kapon
Department of Physics
Technion - Israel Institute of Technology
Haifa, Israel

Supervisor
Prof. Dr. Amit Keren
Department of Physics
Technion - Israel Institute of Technology
Haifa, Israel

ISSN 2190-5053   ISSN 2190-5061 (electronic)
Springer Theses
ISBN 978-3-030-23063-0   ISBN 978-3-030-23061-6 (eBook)
https://doi.org/10.1007/978-3-030-23061-6

This Springer imprint is published by the registered company Springer Nature Switzerland AG
The registered company address is: Gewerbestrasse 11, 6330 Cham, Switzerland

Dedicated to Niman

Supervisor's Foreword

Itzik's thesis deals with two very different projects, but I will focus on the one which is the most innovative and that is the basis of future works in my group.

In normal materials current is proportional to electric field. In superconductors, current is proportional to the vector potential. The proportionality constant is called stiffness and is a quantity of major importance since it provides information on the charge density. Generating a vector potential on a sample without a magnetic field is difficult, therefore measuring stiffness of superconductor is always done in a field, which leads to limitations. To overcome these limitations Itzik uses a very long (60 mm) and narrow (0.25 mm) coil fabricated based on medical technology. Outside of such a coil there is zero magnetic field but a finite vector potential. He then prepares rings from his superconducting material using laser cutters and threads them with his coil. Consequently, the superconductor experiences a vector potential which generates a persistent current going around the ring, without experiencing a magnetic field. The magnetic moment of this current carrying ring is then measured and the stiffness extracted. It turns out that this method is 100 times more sensitive to weak stiffness than the most sensitive method used to date. In a masterpiece of crystal growth, data analysis and applied mathematics, Itzik managed to extract from his data the stiffness tensor.

This extremely sensitive new technique has enabled Itzik to follow the way phase transition occurs in high temperature superconductors very close to the critical temperature T_c. He found that the phase transition occurs in two steps: Initially, supercurrent flows in 2D planes and only at lower temperatures does it begin to flow between planes. This turn of events is not allowed theoretically, and at present I am not aware of a systematic explanation of such a behavior. I am positive that with time Itzik's discovery will draw considerable attention from the superconductivity and statistical physics community.

I consider Itzik's thesis as one of the best works done in my group so far. In the words of one of the referees of the work that was accepted for publication in "Nature Communications": "[This work is] not an incremental finding but a jump in our understanding of cuprate superconductivity."

Haifa, Israel   Prof. Amit Keren
July 2019

Abstract

This thesis consists of two parts. In the first one, we present a new method we have developed to measure the superconducting stiffness tensor $\overline{\rho}_s$, critical current density J^c, and coherence length ξ without subjecting the sample to magnetic field or attaching leads. The method is based on the London equation $\mathbf{J} = -\overline{\rho}_s \mathbf{A}$, where $\mathbf{J}$ is the current density and $\mathbf{A}$ is the vector potential. Using rotor free $\mathbf{A}$ and measuring $\mathbf{J}$ via the magnetic moment of superconducting rings, we extract $\overline{\rho}_s$ at $T \rightarrow T_c$. By increasing $\mathbf{A}$ until the London equation does not hold anymore, we determine J^c and ξ. The technique, named Stiffnessometer, is sensitive to very small stiffness, which translates to penetration depth on the order of a few millimeters. Naturally, the method does not suffer from demagnetization factor complications, the presence of vortices, or out-of-equilibrium conditions. Therefore, the absolute values of the different parameters can be determined.

We apply this method to two different $La_{2-x}Sr_xCuO_4$ (LSCO) rings: one with the current running only in the CuO_2 planes, and another where the current must cross between them. We find different transition temperatures for the two rings, namely there is a temperature range with two-dimensional stiffness. The Stiffnessometer results are accompanied by low-energy μSR measurements on the same sample to determine the stiffness anisotropy at $T < T_c$.

In the second part of the thesis, we investigated whether the spin or charge degrees of freedom were responsible for the nodal gap in underdoped cuprates by performing inelastic neutron scattering and X-ray diffraction measurements on LSCO x = 0.0192. We found that fluctuating incommensurate spin-density wave (SDW) with a bottom part of an hourglass dispersion exists even in this magnetic sample. The strongest component of these fluctuations diminishes at the same temperature where the nodal gap opens. X-ray scattering data from the same crystal show no signature of charge-density wave (CDW). Therefore, we suggest that the nodal gap in the electronic band of this cuprate opens due to fluctuating SDW with no contribution from CDW.

Acknowledgements

I would like to thank my supervisor Prof. Amit Keren, for his excellent guidance, for pushing me relentlessly, and for six years of great science combined with a lot of fun.

Special thanks to Prof. Amit Kanigel for his help and fruitful discussions.

I am grateful to Gil Drachuck for instructing me at the beginning of my way and teaching me valuable lab skills.

I thank Dr. Zaher Salman and Prof. Christof Niedermayer for their great support.

I thank all the students in the group and in the Low-T group: Gil Drachuck, Tom Leviant, Maayan Yaari, Itay Mangel, Nitsan Blau, Alon Yagil, Avior Almoalem, Ilia Khait, Vladi Kalnizki, and Amit Ribak for their friendship and time well spent.

I thank all the technicians and especially Galina Bazalitsky.

Finally, I thank my family for their long-lasting support.

The generous financial help of the Technion is gratefully acknowledged.

Contents

Abbreviations

2D	Two Dimensional
AF	Antiferromagnetism
Ba	Barium
CDW	Charge-Density Wave
Cu	Copper
HTSC	High-T_c Superconductors
La	Lanthanum
LBCO	$La_{2-x}Ba_xCuO_4$
LE-μSR	Low-Energy Muon Spin Rotation
LSCO	$La_{2-x}Sr_xCuO_4$
O	Oxygen
PDE	Partial Differential Equation
SC	Superconductivity
SDW	Spin-Density Wave
SQUID	Superconducting Quantum Interference Device
Sr	Strontium
TSFZ	Traveling Solvent Floating Zone
ZF	Zero Field

Symbols

T_c	Superconducting critical temperature
H_{c1}	First superconducting critical field
H_{c2}	Second superconducting critical field
ρ_s	Superfluid density
x	Strontium doping level of $La_{2-x}Sr_xCuO_4$
Δ	The superconducting gap
T	Temperature
M	Magnetization
D	Demagnetization factor
J	Electric current density
A	Magnetic vector potential
H	External magnetic field
B	Total magnetic field
χ	Magnetic susceptibility
χ_m	Measured susceptibility
χ_0	Intrinsic susceptibility
$H_\perp$	Magnetic field applied perpendicular to the planes
$H_\parallel$	Magnetic field applied parallel to the planes
λ	London penetration depth
ρ_{ab}	In-plane resistivity
ρ_c	Resistivity along the c-axis
q	Neutron momentum transfer
c	Speed of light
e	Electron charge

Chapter 1
Introduction

The existence of two dimensional (2D) superconductivity (SC) in the CuO_2 planes of the cuprates has been demonstrated by either isolated CuO_2 sheets [1, 2], or in bulk, in zero magnetic field [3] or by applying one perpendicular to these planes [4, 5]. In the vicinity of charge stripes formation, the layers are so well decoupled [6] that, in fact, two transition temperatures have been found by resistivity [7] and magnetization in needle shaped samples [8], where the demagnetization factor tends to zero, and the measured susceptibility equals the intrinsic one. The magnetization measurements were done in both c-needles, where the CuO_2 planes are perpendicular to the field direction, and a-needles where the planes are parallel to the field. An updated phase diagram showing the magnetization critical temperature in c-needles T_M^c and a-needles T_M^a is presented in Fig. 1.1. The resistivity critical temperature T_ρ^c of the same samples agrees with T_M^a. The inset shows an example of such magnetization measurement for $La_{2-x}Sr_xCuO_4$ (LSCO) with $x = 0.12$.

However, zero resistivity and diamagnetism do not require bulk superconductivity and can occur due to superconducting islands or filaments. It is not clear whether the observed in-plane superconductivity is a macroscopic phenomenon and if the sample supports global 2D stiffness as expected from Kosterlitz–Thouless–Berezinskii (KTB) theory [9–11]. If it does, there should be a temperature (and doping) range where the intra-plane stiffness $\rho_s^{ab} \equiv 1/\lambda_{ab}^2$ is finite, while the inter-plane stiffness $\rho_s^c \equiv 1/\lambda_c^2$ is zero. Here λ is the penetration depth. To test this, one must be able to measure stiffness very close to the transition temperature.

In light of this, we have developed a new method to measure particularly small SC stiffness without applying external magnetic field. We implemented this technique, called "Stiffnessometer", to investigate 2D SC in LSCO x = 0.125. Thus, the goal of this work is twofold:

1. To present the "Stiffnessometer", the ideas behind it, its operating method and implementation, including data analysis and applications.

© Springer Nature Switzerland AG 2019

I. Kapon, *Searching for 2D Superconductivity in La$_{2-x}$Sr$_x$CuO$_4$ Single Crystals*,
Springer Theses, https://doi.org/10.1007/978-3-030-23061-6_1

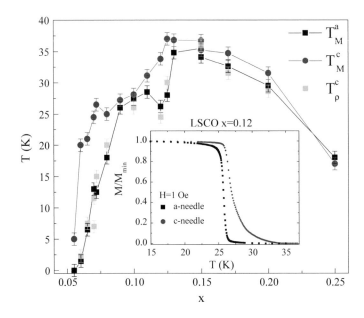

Fig. 1.1 LSCO phase diagram. Temperature versus Sr doping x for a- and c-needles. T_M is the transition temperature taken from magnetization and T_ρ is the one taken from resistivity. The inset introduces an example of magnetization measurement, for two $x = 0.12$ needles at $H = 1$ Oe

2. To scrutinize the possibility of macroscopic 2D superconductivity in the bulk using two different techniques: Low energy muon spin rotation (LE-μSR) and Stiffnessometer. We focus on the "anomalous doping" $x = 1/8$ regime, where the difference between the two transition temperatures is large, and minute inhomogeneity of Strontium doping does not lead to significant deviations in the transition temperatures.

Our major finding is that in LSCO $x = 1/8$, there is a temperature interval of 0.7 K where there is global 2D SC stiffness in the planes, while zero stiffness between them. Namely, in this interval supercurrent can flow in the CuO_2 planes but not between them.

In the rest of this chapter we elaborate on few experimental works showing 2D SC in the cuprates, and give some background on the LSCO compound and experimental techniques. In Chap. 2 we delve into the Stiffnessometer, presenting its capabilities and demonstrate them. Then in Chap. 3 we show our results on LSCO $x = 1/8$.

As a side project, we investigated whether the spin or charge degrees of freedom were responsible for the nodal gap in underdoped cuprates by performing inelastic neutron scattering and X-ray diffraction measurements on LSCO $x = 0.0192$, which is on the edge of the antiferromagnetic phase. We found that fluctuating incommensurate spin-density-wave (SDW) with a bottom part of an hourglass dispersion exists even in this magnetic sample. The strongest component of these fluctuations diminishes at the same temperature where the nodal gap opens. X-ray scattering measurements on the same crystal show no signature of charge-density-wave (CDW). Therefore, we suggest that the nodal gap in the electronic band of this cuprate opens due to fluctuating SDW with no contribution from CDW. This work is presented in Chap. 4.

1.1 Copper Oxides and the $La_{2-x}Sr_xCuO_4$ Compound

High-T_c Superconductivity (HTSC) in the copper oxides (cuprates) was discovered in 1986 [12]. Bendorz and Müller found that the LBCO system had a supercon-ducting transition temperature at $T_c = 30$ K. Later that year, the LSCO compound was discovered, exhibiting superconductivity up to 38 K. All the compounds in the cuprates family have their crystalline structure consists of layers of copper oxide planes, CuO_2, separated by ions of rare earth elements. The spacing between Cu ions is about 3.78 Å. In LSCO, the unit cell consists of two CuO_2 planes, each plane is shifted by half a lattice constant with respect to the other. Between every two CuO_2 planes there are two layers of La(Sr)-O. Figure 1.2 demonstrates the crystalline structure of LSCO, which is the simplest one in the cuprates family.

The cuprates can be doped, either by holes or by electrons, and their charge carrier concentration can be varied. LSCO is hole doped, and its doping mechanism is as follows: The valance of La is 3+ and of O is 2−. Therefore, in the "parent compound" (zero doping, $x = 0$) all the Cu ions are in a Cu^{2+} state; they have one unpaired electron in a d-shell. As x increases, the carrier concentration in the CuO_2 planes is determined by "charge reservoirs" inserted between them. Since the valance of Sr is only 2+, increasing the Sr content by x attracts negative charge from the CuO_2 planes while leaving holes on the Cu sites. Therefore, the hole concentration in LSCO is proportional to the Sr content in the unit cell. Figure 1.3 demonstrates one CuO_2 plane doped with holes.

The cuprates phase diagram is extremely rich, and it is beyond the scope of this thesis to cover all of its complexities. Nonetheless, we will briefly describe its main features of the hole doped part (Fig. 1.4), focusing on LSCO and the relevant phases for our work.

The parent compound is an antiferromagnet (AF) Mott insulator, resulting from strong Coulomb repulsion. The Néel temperature, T_N, is about 300 K. Upon dop-ing, T_N decreases rapidly until it vanishes at $x = 0.02$. However, short range spin order in the form of spin glass or spin density waves remains until $x = 0.12$ inside the SC phase. With extra doping, at $x \approx 0.05$, superconductivity appears, and lasts until $x \approx 0.27$, where regular Fermi liquid appears. The highest transition temperature, $T_c \simeq 38$ K, is achieved at about $x = 0.15$, called accordingly "opti-mal doping". Above and below this doping the compound is referred to as over-doped and underdoped, respectively. Since the T_c versus x curve form a dome-like region, the superconducting phase is often called "superconducting dome". More-over, it is well established that the cuprates order parameter has d-wave symme-try, i.e. it changes sign every 90°. The SC gap function can thus be written as

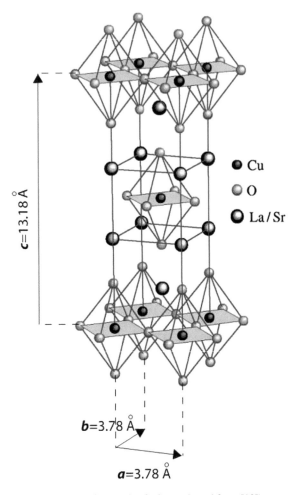

$c=13.18\,\text{Å}$

● Cu
◐ O
◓ La / Sr

$b=3.78\,\text{Å}$

$a=3.78\,\text{Å}$

Fig. 1.2 The crystalline structure of $La_{1-x}Sr_xCuO_4$. (adapted from [13])

$\Delta(k) = (\Delta_0/2)(\cos(k_x a) - \cos(k_y a))$. Another aspect of this symmetry, giving the cuprates even more strange characteristics and making it unconventional SC, is that the order parameter vanishes at four points, "nodes", along the diagonals in the Brillouin zone, giving rise to zero energy excitations.

The normal state of the underdoped and optimally doped sides of the SC dome exhibits peculiar properties, such as high resistivity relative to metals, giving it the name "bad" or "strange" metal. This phase is also characterized by linear temperature dependence of the resistivity, contrary to T^2 behavior expected by Fermi liquid theory. Another hallmark of the bad metal is the absence of quasiparticles. Thus, this regime is referred to as non-Fermi liquid.

Another exotic phase in cuprates is the Pseudogap. This regime starts at T^*, below which a partial gap is opened in the electronic spectrum. This phase is one of the

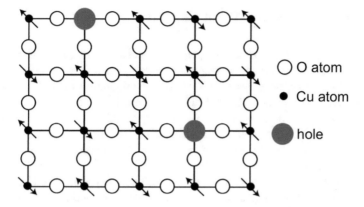

Fig. 1.3 Schematic drawing of CuO_2 plane

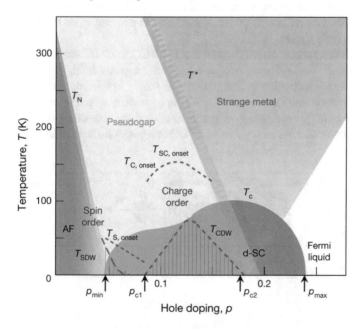

Fig. 1.4 Schematic phase diagram of cuprates. (adapted from [14])

biggest mysteries regarding the cuprates phase diagram, and its origin is strongly debated and is still an open question. One explanation is that this phase stems from the creation of Coopers pairs which are not coherent due to strong phase fluctuations. Another theory relates this regime to other competing orders such as charge and spin density wave.

1.2 2D Superconductivity in Cuprates

Wen et al. [15] investigated the temperature dependence of electrical resistivity, ρ_{ab} and ρ_c, upon application of magnetic fields up to 9 T in single crystals of $La_{2-x}Ba_xCuO_4$ (LBCO) with $x = 0.095$. In the configuration where H was applied perpendicular to the planes, $H_\perp$, the field had a drastic effect on ρ_c, significantly depressing the temperature at which $\rho_c \to 0$, while the effect of $H_\perp$ on ρ_{ab} was rather weak (Fig. 1.5). In contrast, the effect of a parallel applied field was modest for both ρ_{ab} and ρ_c. These results indicate that a two-dimensional phase could exist at high magnetic fields.

Zhong et al. [5] showed similar results in a different cuprate, $La_{1.85}Ca_{1.15}Cu_2O_6$, which has two copper oxide layers in a unit cell. These are summerized in Fig. 1.6, which shows a phase diagram demonstrating how the sample goes through a 2D SC phase before returning to its normal state. The different phases are determined by restitivity measurements under magnetic field perpendicular to the planes, and measuring both ρ_{ab} and ρ_c.

Li and Tranquada et al. [3] showed that in LBCO $x = 0.125$ there is a temperature range in which the layers are decoupled at zero magnetic field. This phenomena is attributed to the formation of charge and spin stripes in this compound.

More experimental evidence of a 2D superconducting phase was presented by Schafgans et al. [16]. A series of magneto-optical reflectance measurements were preformed on underdoped crystals of LSCO at a magnetic field of up to 8 T applied

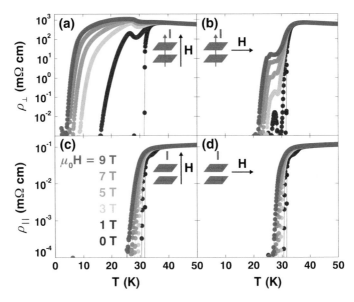

Fig. 1.5 Magnetoresistance in LBCO with $x = 0.095$. Resistivities versus temperature for a range of magnetic fields with the corresponding configurations. Adapted from [15]

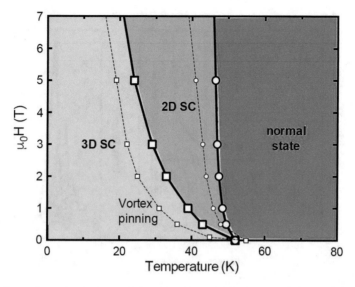

Fig. 1.6 La$_{1.85}$Ca$_{1.15}$Cu$_2$O$_6$ phase diagram. Large squares and circles represent the superconducting transition of the resistivity ρ_c and ρ_{ab} respectively. Magnetic field is applied perpendicular to the planes, and induce decoupling of the layers, thus creating 2D SC region. (adapted from [5])

parallel to the crystal c-axis. These measurements revealed a complete suppression of the interplane coupling, while the in-plane superconducting properties remained intact, suggesting a 2D superconducting state (Fig. 1.7).

Two different sets of theories were conceived to explain the experiments. E. Berg and A. Kivelson proposed a theory which discusses dynamical layer decoupling in stripe-ordered high T_c superconductors [6]. The theory argues that under certain circumstances, the superconducting condensate can occur in a two-dimensional system. This theory was proposed as the underlying cause for the layers decoupling as was observed in LBCO x = 0.125. It was suggested that the existence of stripe order can lead to an enormous suppression of the inter-plane Josephson coupling. This in turn could explain the existence of a broad temperature range in which 2D physics is apparent. Furthermore, Pekker [17] and Vojta [18] proposed independently two complementary theories with the same underlying conclusions. Both theories discuss the different phase transitions in a weakly coupled layered system with c-axis disorder. One prediction of these theories is a temperature region at which an intermediate phase exists where the in-plane superfluid stiffness, ρ_s^{ab}, reaches a finite value while the inter-plane superfluid stiffness ρ_s^c remains zero. Hence, the superfluid splits into an array of 2D puddles with no phase coherence along the c-axis.

Last but not least, we shortly discuss measurements of bulk versus isolated sheets stiffness, emphasizing their different behavior. Hetel et al. [1] measured the superfluid density in underdoped 2-unit-cells-thick Y$_{1-x}$Ca$_x$Ba$_2$Cu$_3$O$_{7-\delta}$ (Ca-YBCO) films using a two-coil mutual inductance method. They found that the stiffness exhibits an abrupt downturn as the temperature approaches T_c, as expected from a 2D KTB

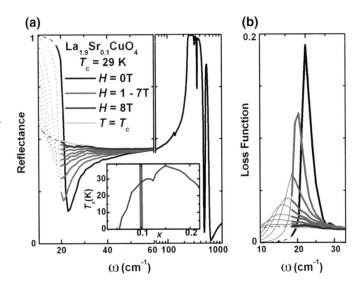

Fig. 1.7 Magneto-optical measurements of LSCO x = 0.1, showing the evolution of the Josephson plasma resonance (JPR) at T = 8 K in magnetic field. The JPR is the only feature in the spectra that is sensitive to field, and by H = 8 T (well below H_{c2}) the JPR signal is suppressed. Adapted from [16]

transition. In contrast, Broun et al. [19] probed the stiffness of bulk underdoped samples of YBCO with width of 0.3 mm using cavity perturbation when H applied perpendicular to the CuO_2 planes. They did not observe 2D KTB behavior, but the stiffness rather went down smoothly through the expected jump in the stiffness.

1.3   Traveling Solvent Floating Zone Method

1.3.1   General Description

Crystal growth using the optical floating zone technique has been extensively used to grow a variety of bulk crystals, particularly of metal oxides such as cuprate superconductors. A large high quality single crystal enables a reliable measurement of physical properties, and is specially important for studying direction dependent ones. High-T_c cuprates superconductors melt incongruently. Namely, they do not melt uniformly and decompose into other substances after solidification, hence growth methods that rely on direct crystallization from self melt are rendered useless for the cuprates. Therefore solution growth have been developed to grow crystals of cuprates. One of the popular methods to grow the high-T_c materials is the Traveling Solvent Floating Zone Method (TSFZ), which allows a high degree of control of the crystal growth parameters.

1.3.2 The Image Furnace

In all image furnaces, the basic concepts is that either ellipsoidal or parabolic mirrors are used to focus light from halogen or xenon lamps onto a vertically held feed rod to produce a molten zone. Figure 1.8 presents a schematic view of the image furnace core parts. The feed and seed material rods are placed inside a quartz tube and mounted on vertical shafts that can be rotated with a variable speed in the same or opposite directions. The quartz tube is used to create a controlled atmosphere, usually of Argon, Nitrogen and Oxygen gas mix, and under desired pressure between 0 and 10 bars. The gap between the two rods is then placed at the common focal point where the temperature can be as high as 3000 °C, which depends on the sample absorption, lamp power, and the applied voltage on the lamps. The high temperature zone melts the rods and creates a molten zone between them. By raising the mirrors and lamps, this zone moves upwards along the feed rod, while the previously molten part is crystalizes after had been moving out ot the high temperature zone. An example of this process is shown in Fig. 1.9.

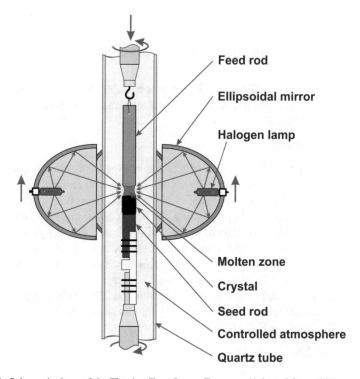

Feed rod

Ellipsoidal mirror

Halogen lamp

Molten zone

Crystal

Seed rod

Controlled atmosphere

Quartz tube

Fig. 1.8 Schematic draw of the Floating Zone Image Furnace. (Adapted from [20])

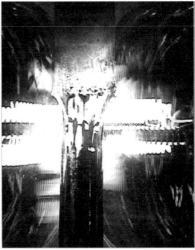

Fig. 1.9 LSCO single crystal during growth. Left: The floating zone furnace at work. Right: Feed and seed rods are connected with a molten zone in between during crystal growth

1.3.3 Key Process Parameters in Crystal Growth

Although very powerful method, growing crystals with the TSFZ is somewhat tidious. It requires carefull optimization of different control parameters in order to achieve a high quality single crystal.

High Quality Feed Rod

The preparation of a feed rod is the initial stage of crystal growth using the TSFZ method. For feed rods made from compacted powder, such as most metal oxides, excess porosity can undermine the stability of the molten zone due to penetration of the melt into the feed rod. This penetration can be attributed to a capillary effect in which the melt is partially absorbed by the cavities among the fine particles in the feed rod. For most materials, such porosity can be decreased by either increasing the pressure at which the rod is compacted or sintering the feed rod at temperatures near its melting point prior to loading it to the image furnace.

Therefore, the feed rod should be homogeneous and uniform in composition and its geometry (straight with constant diameter), and its density should be as close as possible to that of the final single crystal.

Crystallization Rate (Growth Speed)

The growth speed or crystallization rate is unarguably one of the most critical parameters governing crystal quality. Crystallization rate can strongly vary from 240 mm/h (GaAs) to 0.05 mm/h (Bi-based superconductors). It has been widely reported thatchanging growth speed can affect the grown crystal in terms of crystal size, formation of bubbles, cracks, chemical composition, crystal alignment, twin formation and has a great influence on the solid-liquid interface and molten zone stability. The growth rate is mainly restricted by the slow solution diffusion process at the solid-liquid interface boundary, thus the typical growth rate required for optimal crystal quality depends on whether the materials melt congruently or incongruently. For congruently melting materials the composition of the molten zone is the same as the feed rod, crystallization process is not much limited by slow diffusion processes and relatively higher growth speed can be achieved. For incongruently melting materials, the composition of the melt differs from the one of the original solid. This difference necessitates solution diffusion at the solid-liquid interface, which generally takes place slowly and therefore limits growth speed to a very slow rate.

Growth Atmosphere and Gas Pressure

Both atmosphere and gas pressure are crucial parameters when growing crystal in the TSFZ method and play a key difference between success and failure. These parameters are fairly easily controlled during the crystal growth by selecting the right gas mixture coming in, and the desired pressure coming out from the quartz tube. The main reason quoted for growing in higher than atmospheric pressure is to reduce the vaporization of volatile components from the sample.

Lamp Power and Temperature of the Molten Zone

The power level depends mainly on the chemical properties of the grown material, but is also affected by factors such as gas content and pressure, growth rate, density and diameter of the feed rod, lamp de-focusing and the temperature gradient around the molten zone. For incongruently melting materials it is extremely important to adjust the power level according to the material phase diagram, then it must be kept constant. Failing to fulfill this condition will result in either secondary phases or completely wrong chemical composition of the grown crystal.

1.3.4 $La_{2-x}Sr_xCuO_4$ *Crystal Growth*

Powders of CuO (99.9%), La_2O_3 (99.99%) and $SrCO_3$ (99.9%) were dried at high temperature between 500 to 1050 °C, then weighted accordingly to the calculated stoichiometric values. The desired doping level of the end product crystal is therefore determined in the beginning of the process by adding the right amount of $SrCO_3$ into the mixture. An extra 2.5% of CuO were added due to evaporation during the crystal growth. The weighted powders were mixed and grinded together until smooth and homogeneous texture is achieved. The mixed powder was placed in alumina crucible and underwent firing at 960 °C in a box furnace. Such high temperatures induce diffusion of the reactants and binds the chemicals together to form $La_{2-x}Sr_xCuO_4$. The grinding and firing process was repeated three time in order to eliminate possible impurity phases. After this process was completed, the powders were inspected with x-ray analysis to ensure right doping concentration and purity. The second stage of preparation involves making a cylindrical shaped rod, which will be used as a feed and seed for the crystal. The powder mixture was compacted into a rubber tube which was then inserted into an isostatic press. Pressure of up to 60,000 psi (4000 bar) compacts the powder into a long rod (up to 20 cm), reaching approximately 60% of the crystal density. The compacted rod was sintered at $T = 1230\,^\circ\text{C}$ near its melting point temperature for 24 h. This step brings its density very close ($\approx 95\%$) to the crystal one. Figure 1.10 demonstrates such feed rod.

The crystals were grown under elevated pressure of mixed Argon/Oxygen (10:1) atmosphere. A slow rate of 1 mm/hour was chosen to let the diffusion in the melt to take place, the liquid and the solid being of different composition. The feed and seed/crystal were rotated in opposite directions at 15 RPM in order to improve the liquid homogeneity. All the growths were ended voluntarily after the whole feed rod was consumed by growth, yielding black semi-metallic color crystal with lengths ranging from 70 to 120 mm and typical diameter of 5–7 mm depending on the starting rod dimensions and pull rate of the feed rod. An exemplary crystal is shown in Fig. 1.11. After growth, the crystals were annealed in Argon atmosphere at 850 °C for 120 h to remove excess oxygen and relieve thermal stress.

Fig. 1.10 A sintered feed rod of LSCO

Fig. 1.11 As grown single crystal of LSCO

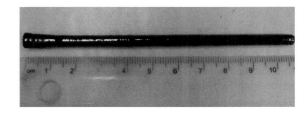

1.4 SQUID Magnetometer

Superconducting Quantum Interference Devices (SQUIDs) are commonly used to detect the smallest magnetic signals and function as highly sensitive magnetic flux-to-voltage transducers. The SQUID relies on the physical principal of the Josephson junction, making it sensitive to a change in magnetic flux of one flux quantum $\Phi_0 = \frac{h}{2e} = 2.07 \times 10^{-7}\ G\,cm^2$.

The measurement system that was used in this work was an S600 SQUID SUSCEPTOMETER of CRYOGENIC LTD. This system can work at either high magnetic field regime up to 6.5 T, or at a low field regime up to 200 G, with field resolution of 0.001 G. The measurement is performed by moving a sample trough a set of pickup coils in a configuration known as a second-order Gradiometer. The movement induces a change of magnetic flux and creates screening currents that flow into the flux transformer. This flux change is detected by the SQUID device. The output voltage is converted to physical units of magnetic moment. More information regarding the SQUID operation is discussed in Chaps. 2 and 3 where it is relevant.

1.5 Muon Spin Rotation (μSR)

Muon spin rotation, relaxation or resonance (μSR) is a technique used for studying magnetic properties in materials, by directly measuring the time dependence of the muons spin, after injecting it into the sample. This technique allows detection and measurement of static magnetic fields in the range of $10^{-5} - 1$ Tesla or magnetic field fluctuations on a time scale of $10^{-3} - 10^{-11}$ s.

1.5.1 Principle of μSR

The production of muons is achieved by using high energy proton beams produced in cyclotrons. The protons are fired into a target to produce pions via

$$p + p \rightarrow \pi^+ + p + n$$

and then the pions decay into muons via

$$\pi^+ \rightarrow \mu^+ + \nu_\mu$$

The pions that decay at the target surface have zero momentum so the outgoing muon and neutrino will have opposite momenta. The pions are spin-less particles, thus by angular momentum conservation the muon and the neutrino must have opposite spins. The neutrino has a definite chirality, and its spin must always be aligned in the opposite direction to its momentum, implying that the muons produced in this process are 100% spins polarized. The muons are transported through the beam-line using a system of magnets, that maintain the spin polarization. They hit the sample with an energy of 4 MeV, where they lose their energy via Coulombic scattering processes which have no effect on their spin, leaving the spins polarized in a certain direction.

In the presence of external or internal magnetic field, the muon spin begins to rotate with an angular frequency of $\omega_\mu = \gamma_\mu B$, where γ_μ, the gyro-magnetic ratio, is 13.55 KHz/Gauss. The life time of the muon is $\tau_\mu = 2.2\,\mu$s, and it decays in a three body process

$$\mu^+ \rightarrow e^+ + \nu_e + \bar{\nu}_e$$

The angular distribution of the emitted positrons is shown in Fig. 1.12. The decay involves the weak interaction which violates parity, and leads to the positron being emitted preferentially along the direction of the muon's spin after it has been decayed. This effect allows one to measure the polarization of precessing muons.

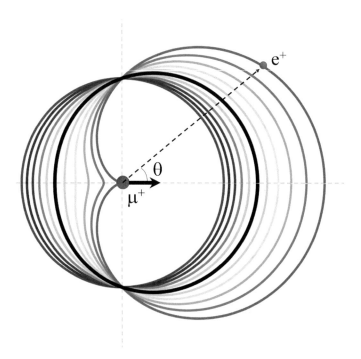

Fig. 1.12 Illustration of the angular distribution of emitted positrons from muon decay. Black arrow represents the muon spin. Different curves represent different positron energies, the higher (lower) one is in red (purple). The black curve is the angular distribution averaged over all positron energies

1.5.2 The μSR Experiment

A schematic diagram of the μSR experiment is shown in Fig. 1.13. Once a muon is implanted in the sample, a clock starts running and stops only when a positron is detected, using an array of plastic scintillators connected via a light-guide to a set of photo-multipliers. In the simplest setup we have two detectors, the detector in the incident direction of the muon named the "forward detector" and the "backward detector" in the opposite direction.

In principle there could be up to six detectors: Forward and backward, right and left, top and bottom. From each detector a histogram of decay times is built (e.g. N_F and N_B for the forward and backward detectors, respectively). The number of

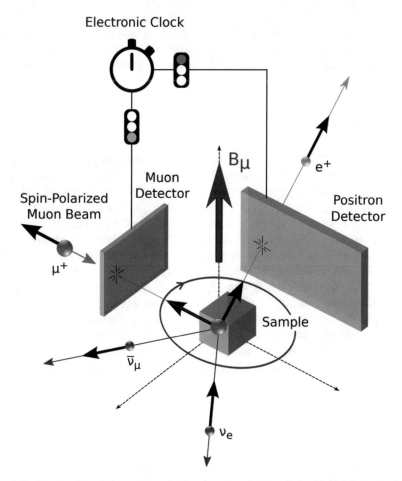

Fig. 1.13 Sketch of a μSR experiment in the transverse-field configuration. Black and colored arrows represent the particles spin and momentum, respectively. Drawing by Andreas Suter (PSI)

positrons detected in all the detectors decays exponentially with time, but if the muon
"feels" a magnetic field, the histogram will oscillate with frequency ω_μ. The muon
polarization can be extracted from the normalized difference between the histograms
of the forward and backward detectors:

$$A(t) = \frac{N_F(t) - N_B(t)}{N_F(t) + N_B(t)}$$

A(t) is known as the asymmetry. It contains information about the local magnetic
environment in materials, and is usually measured as a function of temperature.
When no external field is applied, the muon will precess and reveal the internal
field distribution. This method is known as Zero Field μSR (ZF-μSR) and it is
useful in research of ferromagnets, anti-ferromagnets and spin-glasses. μSR can
also be carried out when applying external magnetic field. There are two different
configurations for μSR in external field: Longitudinal field μSR, in which the initial
muon spin polarization is parallel to the applied field direction, and transverse field
μSR, in which the initial muon spin polarization is perpendicular to the applied field
direction.

1.5.3   Low Energy μSR

In Low energy μSR (LE-μSR), by controlling the muons energy E, the muons stop
with high probability at some chosen depth inside the sample while keeping their
polarization intact. The experiment was performed in the LEM beam line [21] at
the Swiss Muon Source SμS, Paul Scherrer Institute, Villigen, Switzerland. 4 MeV
spin-polarized muons are stopped at a moderator, made of 300 nm thick layer of solid
Argon grown on top of a silver foil. They are then accelerated to a chosen energy
between 1 and 30 keV by applying a voltage difference between the foil and the
sample. The sample holder is placed on a sapphire plate hence electrically isolated.

The whole chamber is kept in ultra high vacuum of 10^{-10} mbar, and the stopping
and accelerating processes of the muons preserve most of the polarization. Once in
the sample, the muon spin rotates in the local external or internal magnetic field and
the time dependent polarization is reconstructed from asymmetry in the positrons
decay, which are emitted preferentially in the muon spin direction.

More information, including data analysis methods, is presented in the main text
in Chap. 3.

References

1. Hetel I, Lemberger TR, Randeria M (2007) Quantum critical behaviour in the superfluid density
   of strongly underdoped ultrathin copper oxide films. Nat Phys 3:700
2. Božović I, He X, Wu J, Bollinger AT (2016) Dependence of the critical temperature in over-
   doped copper oxides on superfluid density. Nature 536:309. https://www.nature.com/articles/
   nature19061

3. Li Q, Hücker M, Gu GD, Tsvelik AM, Tranquada JM (2007) Two-dimensional superconducting fluctuations in stripe-ordered $La_{1.875}Ba_{0.125}CuO_4$. Phys Rev Lett 99:067001. https://journals. aps.org/prl/abstract/10.1103/PhysRevLett.99.067001

4. Baity PG, Shi X, Shi Z, Benfatto L, Popović D (2016) Effective two-dimensional thickness for the berezinskii-kosterlitz-thouless-like transition in a highly underdoped $La_{2-x}Sr_xCuO_4$. Phys Rev B 93:024519

5. Zhong R, Schneeloch JA, Chi H, Li Q, Gu G, Tranquada JM (2018) Evidence for magnetic-field-induced decoupling of superconducting bilayers in $La_{2-x}Ca_{1+x}Cu_2O_6$. Phys Rev B 97:134520

6. Berg E, Fradkin E, Kim E-A, Kivelson SA, Oganesyan V, Tranquada JM, Zhang SC (2007) Dynamical layer decoupling in a stripe-ordered high-T_c superconductor. Phys Rev Lett 99:127003

7. Tee XY, Ito T, Ushiyama T, Tomioka Y, Martin I, Panagopoulos C (2017) Two superconducting transitions in single-crystal $La_{2-x}Ba_xCuO_4$. Phys Rev B 95:054516

8. Drachuck G, Shay M, Bazalitsky G, Berger J, Keren A (2012) Parallel and perpendicular susceptibility above T_c in $La_{2-x}Sr_xCuO_4$ single crystals. Phys Rev B 85:184518

9. Kosterlitz JM, Thouless DJ (1973) Ordering, metastability and phase transitions in two-dimensional systems. J Phys C Solid State Phys 6:1181

10. Berezinskii VL (1972) Destruction of long-range order in one-dimensional and two-dimensional systems possessing a continuous symmetry group. II. Quantum systems. Sov J Exp Theor Phys 34:610

11. Kosterlitz JM (1974) The critical properties of the two-dimensional XY model. J Phys C Solid State Phys 7

12. Bednorz JG, Müller KA (1986) Possible high T_c superconductivity in the Ba–La–Cu–O. Z Phys B Cond Mat 64

13. Damascelli A, Hussain Z, Shen Z-X (2003) Angle-resolved photoemission studies of the cuprate superconductors. Rev Mod Phys 75:473. https://doi.org/10.1103/RevModPhys.75.473

14. Bernhard K, Kivelson SA, Norman MR, Uchida S, Zaanen J (2015) From quantum matter to high-temperature superconductivity in copper oxides. Nature 518

15. Wen J, Jie Q, Li Q, Hücker M, Zimmermann MV, Han SJ, Xu Z, Singh DK, Konik RM, Zhang L, Gu G, Tranquada JM (2012) Uniaxial linear resistivity of superconducting $La_{1.905}Ba_{0.095}CuO_4$ induced by an external magnetic field. Phys Rev B 85:134513

16. Schafgans AA, LaForge AD, Dordevic SV, Qazilbash MM, Padilla WJ, Burch KS, Li ZQ, Komiya S, Ando Y, Basov DN (2010) Towards a two-dimensional superconducting state of $La_{2-x}Sr_xCuO_4$ in a moderate external magnetic field. Phys Rev Lett 104:157002

17. Pekker D, Refael G, Demler E (2010) Finding the elusive sliding phase in the superfluid-normal phase transition smeared by c-axis disorder. Phys Rev Lett 105:085302

18. Mohan P, Goldbart PM, Narayanan R, Toner J, Vojta T (2010) Anomalously elastic intermediate phase in randomly layered superfluids, superconductors, and planar magnets. Phys Rev Lett 105:085301

19. Broun DM, Huttema WA, Turner PJ, Özcan S, Morgan B, Liang R, Hardy WN, Bonn DA (2007) Superfluid density in a highly underdoped $YBa_2Cu_3O_{6+y}$ superconductor. Phys Rev Lett 99:237003. https://doi.org/10.1103/PhysRevLett.105.085301

20. Guo-Yong Z, Mitch C, Cheng-Tian L (2017) $(Li_{1-x}Fe_x)OHFeSe$ superconductors: crystal growth, structure, and electromagnetic properties. Crystals 7:167

21. Prokscha T, Morenzoni E, Deiters K, Foroughi F, George D, Kobler R, Suter A, Vrankovic V (2008) The new μE4 beam at PSI: a hybrid-type large acceptance channel for the generation of a high intensity surface-muon beam. Nucl Instrum Methods Phys Res Sect A Accel Spectrom Detect Assoc Equip 595:317–331

Chapter 2
Stiffnessometer, a Magnetic-Field-Free Superconducting Stiffness Meter and Its Application

In this chapter, we introduce a new method to measure superconducting stiffness ρ_s, critical current density J^c, and coherence length ξ without subjecting the sample to magnetic field or attaching leads. We start by presenting the technique and the underlying ideas behind it, then show data taken on different types of superconductors (type I, II and high T_c superconductor), discuss the strengths and weaknesses of the method, and finally demonstrate its application to LSCO x = 0.12. We put stress on verification experimental tests taken to assure that our method works.

Superconducting stiffness ρ_s is defined via the quantum mechanical, gauge invariant relation between the current density $\mathbf{J}$, the vector potential $\mathbf{A}$, and the complex order parameter $\psi = |\psi|\, e^{i\varphi(\mathbf{r})}$ according to

$$\mathbf{J} = \rho_s \left(\frac{\hbar c}{q} \nabla \varphi - \mathbf{A} \right) \tag{2.1}$$

where q is the carriers charge [1]. When $\nabla \varphi = 0$[1] the London equation is obtained:

$$\mathbf{J} = -\rho_s \mathbf{A}. \tag{2.2}$$

ρ_s can be expressed in units of length via

$$\rho_s = \frac{c}{4\pi\lambda^2}, \tag{2.3}$$

where λ is known as the penetration depth, and describes the length upon which external magnetic field decays into a superconductor, namely the Meissner effect.

---

[1]The conditions under which this assumption holds are discussed later in this chapter and on the next one.

© Springer Nature Switzerland AG 2019
I. Kapon, *Searching for 2D Superconductivity in La2−xSrxCuO4 Single Crystals*,
Springer Theses, https://doi.org/10.1007/978-3-030-23061-6_2

London's equation can be derived from the following argument: Quantum mechanically, the canonical momentum operator obeys $\mathbf{p} = m\mathbf{v} + q\mathbf{A}/c$. Bloch's theorem states that $\langle \mathbf{p} \rangle$ must vanish at the ground state. Together with the current density definition $\mathbf{J} = n_s q \mathbf{v}$, one gets the London equation $\mathbf{J} = -n_s q^2 \mathbf{A}/mc$.

Since ρ_s provides information on the ratio between superfluid density and effective mass, it is the most fundamental property in the study of superconductors. For example, in underdoped high temperature superconductors (HTSC) the transition temperature T_c is found to be proportional to the stiffness at low temperatures. This finding is known as the Uemura relation [2]. It has also been found by Homes et al. [3] that in the cuprates T_c multiplied by the DC conductivity just above T_c is proportional to the superfluid stiffness. These two scaling laws must play a key role in any theory of HTSC [2]. Another important example is that the temperature dependence of the stiffness, $\rho_s(T)$, is determined by the symmetry of the superconducting order parameter, thus allows one to infer the underlying pairing mechanism, e.g. s-wave or d-wave [4].

The standard method to measure the stiffness is via the penetration depth; one applies a magnetic field and measures its penetration into a material [2, 4–6]. However, magnetic field raises issues one must consider: first, it is essential to take into account the sample shape via the concept of the demagnetization factor. This factor is known exactly only for ellipsoidal samples, which are nearly impossible to come by. Second, magnetic fields introduce vortices, which can complicate the interpretation of the penetration depth measurements. Third, all methods have an inherent length scale window. The longest penetration depth that has been measured to the best of our knowledge is $10\,\mu$m [4–6]. This is far shorter than a typical sample size. Therefore, there is a temperature range below T_c at which $\lambda > 10\,\mu$m where the behavior of ρ_s is obscured. For highly anisotropic samples, this range could extend to temperatures well below T_c.

Similarly, there is no direct way to measure J_c. The standard method is to connect leads, and to determine the current at which voltage develops across the sample. However, the voltage first develops in the vicinity of the leads, where heat is generated. This heat affects the rest of the sample.

Here we present a new instrument to measure stiffness in zero magnetic field based on the London equation. This method determines ρ_s directly without the use of the penetration depth concept. Consequently, we name the instrument Stiffnessometer. As we explain below, the Stiffnessometer can measure very weak stiffness, which corresponds to λ ranging from tens of microns to few millimeters. This allows measurements of stiffness closer to the critical temperature T_c than ever before, or measuring the stiffness of very anisotropic systems. Finally, vortices or demagnetization factor are not a problem for the Stiffnessometer since the measurement is done in zero field. We also demonstrate that the Stiffnessometer can measure critical currents without leads and hence provide information on the coherence length ξ.

2.1 Experimental Setup

The method is based on the fact that outside an infinitely long coil (defining the $\hat{z}$ direction), the magnetic field is zero while the vector potential is finite. This vector potential is tangential and points in the $\hat{\varphi}$ direction. When such an inner-coil is placed in the center of a superconducting ring, the vector potential leads to a current density in the ring according to Eq. 2.1. This current flows around the ring and generates a magnetic moment. The magnetic moment is detected by moving the ring and the inner-coil rigidly relative to a pickup-loop. The concept of the measurement is depicted in Fig. 2.1a. A typical inner-coil and two superconducting rings of the cuprate superconductor $La_{2-x}Sr_xCuO_4$ (LSCO) with $x = 0.12$ are shown in

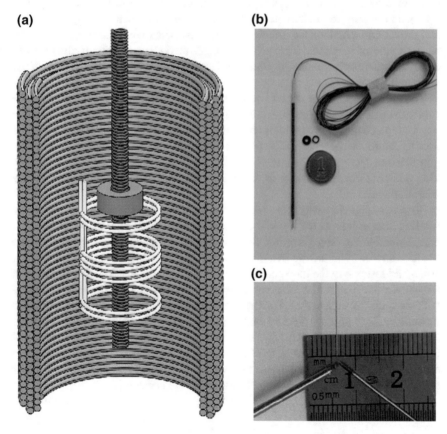

Fig. 2.1 Experimental setup. **a** An illustration of the Stiffnessometer: The superconducting ring is threaded by an inner-coil, placed in the center of a gradiometer, and surrounded by a main-coil that serves as a shim coil. **b** A typical inner-coil, 60 mm long with 2 mm outer diameter. The data collected in this work are taken using this inner-coil. Also shown are two $La_{2-x}Sr_xCuO_4$ rings with a rectangular cross-section **c** A zoom-in on other inner-coils with outer diameters ranging from 2 to 0.25 mm, and length of 60 mm

Fig. 2.1b. More details on these coil and rings are given in the analysis section. In Fig. 2.1c we present a zoom-in on three different coils with outer diameters of 2, 0.8, and 0.25 mm. They have two layers of wires, and their length is 60 mm. Our Stiffnessometer is an add-on to a Cryogenic SQUID magnetometer.

Rather than using a single pickup-loop we actually use a second order gradiometer. It is made of three winding groups. The outer two are made of two loops each wound clockwise and the inner group is made of 4 loops wound anticlockwise. This is also demonstrated in Fig 2.1a. The gradiometer ensures that a magnetic moment generates a voltage only when it is in the vicinity of the gradiometer center. Also, any field uniform in space gives zero signal even if it drifts in time.

The gradiometer is connected to a superconducting quantum interference device (SQUID). The output voltage V of the device is proportional to the difference between flux threading the different loops of the gradiometer. We record $V(z)$ as the relative distance between the gradiometer and the ring changes when the ring and inner-coil move. Our gradiometer detects magnetic moments within a range of 15 mm on each side of its center. This sets the length of our inner-coils. It allows us to detect contribution from the ring while minimizing contribution from the ends of the inner-coil.

The measurements are done in zero gauge field cooling (ZGFC) procedure: we cool the ring to a temperature below T_c, turn on the current in the inner-coil when the ring is superconducting, and measure while warming. A measurement above T_c is used to determine the value of the flux inside the inner-coil. For any circle of radius r in the ring $\nabla \varphi = l/r \, \hat{\varphi}$ where l is an integer. The ZGFC procedure sets $l = 0$. This value of l does not change as $\mathbf{A}$ is turned on, and Eq. 2.2 holds throughout the measurements.

To better appreciate why $\nabla \varphi = 0$, one can view the phase φ as an in-plane arrow. Cooling at $\mathbf{A} = 0$ must set $\nabla \varphi = 0$ to minimize the kinetic energy, namely, all the arrows point in the same direction. Since the phase is quantized, to change φ means to make a twist of all arrows in a closed loop, such that the phase between the first arrow and last one in the loop changes by 2π. This would lead to a discontinuity in the phase value, a procedure that costs energy. A nice analog is a ferromagnetic ring with the spins pointing in the same direction. Rotating the last spin with respect to the first one by 2π requires to break a bond. This procedure is not energetically favorable for a ferromagnet (or the SC ring). Therefore, when turning $\mathbf{A}$ on after cooling, all the arrows continue to point in the same direction and $\nabla \varphi = 0$, until A exceeds a critical value. At this point, the current in the SC is too high and it is worthwhile for the superconductor to "break a bond" and reduce the current.

2.2 Results

A typical measurement is demonstrated in Fig. 2.2. The red symbols represent the signal when the entire inner-coil has moved through the pickup-coil at $T > T_c$. Before the lower-end of the inner-coil has reached the gradiometer, the flux through it is

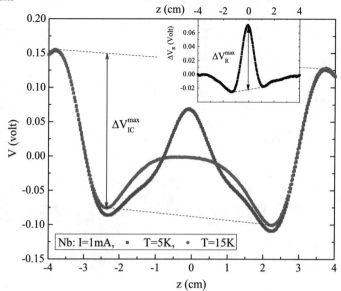

Fig. 2.2   Raw data. SQUID signal for a Nb ring at high temperature when the ring is not superconducting and at low temperature when the ring is superconducting. The inset shows the difference
between them

zero. During the time the lower-end of the inner-coil transverses the gradiometer, its
contribution to the total flux changes from zero to positive, then to negative and back
to zero again. The upper-end of the inner-coil has the opposite effect; its contribution
to the flux goes from zero to negative to positive and back to zero. But there is a time
(or distance) delay between the lower-end and upper-end contributions, leading to the
observed signal. A linear drift of the voltage can be easily evaluated as demonstrated
by the dotted lines. We define the inner-coil maximum voltage difference ΔV_{IC}^{max} as
demonstrated in Fig. 2.2.

At $T < T_c$ the ring adds its own signal, as shown in Fig. 2.2 by blue symbols.
The ring produces current that generates opposite flux to the one in the inner-coil.
The ring signal is concentrated on a narrower range on the z axis. By subtracting
the high temperature measurement from the low temperature one, it is possible to
obtain the signal from the ring alone ΔV_R as demonstrated in the inset of Fig. 2.2.
We define the maximum ring voltage difference ΔV_R^{max} as shown in the inset. The
ratio $\Delta V_R^{max}/\Delta V_{IC}^{max}$ stores the information on the stiffness, as will be discussed
in the Data Analysis subsection. For a given inner-coil and current it is enough to
determine V_{IC} once.

In Fig. 2.3 we present the Stiffnessometer signal evolution with temperature from
an LSCO $x = 0.12$ ring. At temperatures between 5 and 22 K there is no change in
the signal. But, between 22.5 K and $T_c = 27.9$ K the signal diminishes rapidly, as
expected. The normalized voltage difference $\Delta V_R^{max}/\Delta V_{IC}^{max}$ is plotted in the inset
of Fig. 2.3.

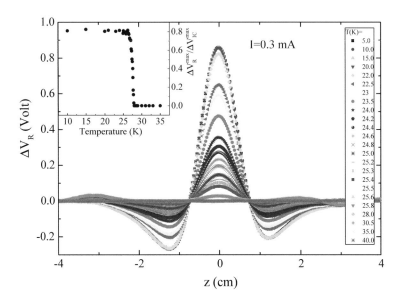

Fig. 2.3 Temperature dependence. The SQUID signal difference between high and low temperatures ΔV_R for an LSCO $x = 0.12$ ring at different temperatures. The CuO$_2$ planes are perpendicular to the ring symmetry axis. The inset shows the normalized voltage difference as a function of temperature

There is a risk that field generated in the inner-coil leaks since no coil is infinitely long or perfect. To overcome this leak, a main coil, also shown in Fig. 2.1, acts as a shim to cancel the field on the ring when it is at the gradiometer center. Our main-coil has a field resolution of 10^{-3} Oe from 0 up to 200 Oe. Therefore, we can keep the field on the ring as low as 1 mOe.

To ensure that our signal is not due to leakage of magnetic field from the inner-coil or any other field source, we perform two tests. In the first one we apply current in the inner-coil, measure the field leakage at the ring position using an open ring and cancel it using the main coil. Then we increase the field by only 1 mOe. The measurements before and after the field increase are depicted in Fig. 2.4a. They indicate that we can cancel the field in the ring position to better than 1 mOe. Clearly in zero field there is no signal. In the second test we measure the stiffness (zero field and applied current in the inner-coil) of closed and open rings, which are otherwise identical in size. The results are shown in Fig. 2.4b. The signal from a closed ring is much bigger than the background from an open one. In Fig. 2.4c we repeat this measurement with an applied field in the main coil of 1 Oe, and no current in the inner-coil. In this case both open and closed rings give strong and similar signal. The difference between the two signals is consistent with the missing mass in the open ring. These tests confirm that the field leakage is not relevant to our stiffness measurement. Our ability to determine small stiffness depends on how well we can cancel the field at the ring position.

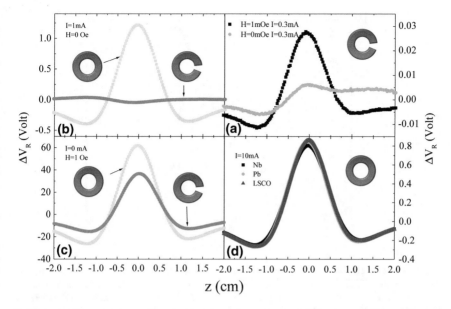

Fig. 2.4 Experimental tests. **a** The signal with a current of 0.3 mA in the inner-coil and fields of 0 and 1 mOe demonstrating the quality of the field cancelling procedure. **b** The SQUID signal for an open and closed rings when the field is zero and the vector potential is finite. **c** A test experiment: the SQUID signal for an open and closed rings when the vector potential is zero but the field is finite. **d** Demonstrating that when λ is much smaller than the sample size the signal is material independent

Another important test of the Stiffnessometer comes from comparing the signal from rings of exactly the same dimensions, but made from different materials. At temperatures well below T_c the stiffness is expected to be strong, namely, the penetration depth should be much shorter than all the ring dimensions. In this case, a superconducting ring produces a current which exactly cancels the applied flux through it, regardless of the material used. Therefore, all materials should produce the same signal. This is demonstrated in Fig. 2.4d for Niobium (Nb), Lead (Pb) and LSCO. They all have the same ΔV_R.

The Stiffnessometer can also be used to measure critical currents. This is depicted in Fig. 2.5 for the LSCO ring at $T = 25.8$ K. The signal from the ring grows linearly with the applied current in the inner-coil until ΔV_R^{max} reaches a saturation value. It means that the superconductor can generate only a finite amount of current. Therefore, we are detecting a critical current, but in a thermal equilibrium fashion since we do not use leads or inject power into the system as usually done in critical current measurements [7, 8]. Thus, it is more adequate to compare our measurements of critical current with theoretical expectations. Moreover, the critical current density in the ring must be related to the applied critical current divided by dimensions of area, which in our case are of order 1 mm^2. This demonstrates that the critical

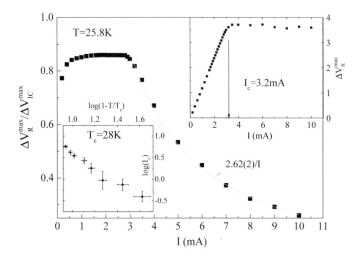

Fig. 2.5 Critical current. The normalized voltage difference, which is proportional to the stiffness, induced by an LSCO ring as a function of the current in the inner-coil. The data were taken at $T = 25.8$ K. The stiffness is practically current independent up to I_c and then falls off like $1/I$ as demonstrated by the solid red line. The upper right inset shows the unnormalized signal. The lower left inset demonstrates that the critical current is proportional to $(T - T_c)^{1.75}$

current we are measuring must be on the order of 1 Acm^{-2}; a more detailed analysis is given below. The critical current density measured with the Stiffnessometer is several orders of magnitude smaller than the critical current density measured by other methods, which are on the order of 10^6 Acm^{-2}.

As the current in the coil exceeds I_c, vortices start to flow into the center of the ring, so that the current density in the ring never exceeds the critical value J_c. In other words $\nabla \varphi = l/r$ with $l \neq 0$. Therefore, for $I > I_c$, the current in the ring and ΔV_R^{max} are fixed. In contrast, ΔV_{IC}^{max} increases linearly with I so that the stiffness decreases like $1/I$. This behavior is demonstrated in the upper right inset and main panel of Fig. 2.5.

2.3 Data Analysis

Before analyzing the Stiffnessometer signal it is essential to determine the realistic vector potential generated by our inner-coil. The vector potential outside of an infinitely long coil is given by

$$\mathbf{A}_{IC} = \frac{\Phi_{IC}}{2\pi r}\hat{\varphi}, \tag{2.4}$$

where r is the distance from the center of the coil, and Φ_{IC} is the flux produced by the inner-coil. To check the validity of this expression in our case we calculated numerically the magnetic field B_z and vector potential A_φ (in the Coulomb gauge) produced by the inner-coil as a function of r and z. This coil is 6.0 cm long, has inner diameter (I.D.) of 0.08 cm, outer diameter (O.D.) of 0.2 cm, 4 layers, and 1600 turns in total. In the calculation we used a current of 1 A. The measured LSCO ring has an I.D. of 0.2 cm, O.D. of 0.5 cm, and height of 0.1 cm. Figure 2.6 shows the result of the calculations. The $1/r$ approximation, presented by the solid line, is perfect for our ring size and even for much larger rings. The calculation also shows that the strongest field just outside of the inner-coil is 10^3 times smaller than the field at its center.

Analyzing the Stiffnessometer signal is done in two steps and on two levels. The steps are first to consider a single pickup-loop and only then a gradiometer. The levels are: weak stiffness and strong stiffness. Weak stiffness means that the vector potential on the ring is only due to the applied current. The vector potential generated by the internal current of the ring is ignored. This approximation is valid when the ring current is small, namely, the stiffness is weak, or the penetration length is longer than the sample dimensions. The weak stiffness analysis is analytical and valid close to T_c. At the strong stiffness level the self vector potential is taken into account. This leads to a partial differential equation (PDE), which we solve numerically.

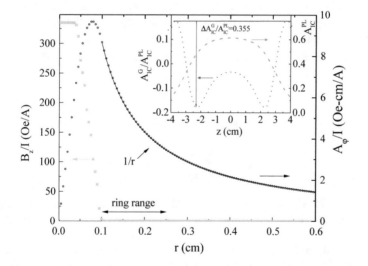

Fig. 2.6 Vector potential and magnetic field profile. Numerical calculation of the vector potential and magnetic field per current at $z=0$ for one of our inner coils. The coil parameters are: current $I = 1$ A, length $l = 6.0$ cm, inner diameter $= 0.08$ cm, outer diameter $= 0.2$ cm, there are 4 layers and 1600 turns. The ring position relative to the inner-coil center is demonstrated by the arrows. The vector potential falls off like $1/r$ over the range of the ring as the solid line demonstrates. Inset: A_{IC}^{PL} and A_{IC}^{G}/A_{IC}^{PL} as a function of z, as explained in the main text

2.3.1 Single Pickup-Loop

Had we used a single pickup-loop, the voltage would have been proportional to the flux threading it $\int \mathbf{B}d\mathbf{a} = \int \mathbf{A}d\mathbf{l} = 2\pi R_{PL}A(R_{PL})$, where $R_{PL} = 1.3$ cm stands for our pickup-loop radius. Above T_c, maximum voltage is achieved when the pickup-loop is at the center of the inner-coil so that $V_{IC}^{max} = k\Phi_{IC}$ where k is a proportionality constant. Similarly, a ring at the center of and parallel with a pickup-loop would have generated a maximum voltage proportional to its own flux, namely, $V_R^{max} = k2\pi R_{PL}A_R(R_{PL})$ where A_R is the vector potential generated by the ring. Therefore,

$$\frac{\Delta V_R^{max}}{\Delta V_{IC}^{max}} = \frac{A_R(R_{PL})}{A_{IC}(R_{PL})} \tag{2.5}$$

so we only need to calculate the ratio of the vector potentials at the pickup-loop radius.

2.3.1.1 Weak Stiffness

The current from each ring element is $dI(r) = J(r)hdr$ where h is the ring height and dr is a ring element width. Using the London equation, the magnetic moment generated by each ring element is $dm = \frac{\pi r^2}{c}dI = \frac{r\rho_s\Phi_{IC}h}{2c}dr$. Integrating from the inner to the outer radii yields the total moment of the ring $m = \frac{\rho_s\Phi_{IC}h}{4c}(R_{out}^2 - R_{in}^2)$, and

$$A_R = \frac{m}{r^2} \tag{2.6}$$

Using Eq. 2.3, the penetration depth is given by

$$\lambda^2 = \frac{h(R_{out}^2 - R_{in}^2)}{8R_{PL}}\frac{A_{IC}(R_{PL})}{A_R(R_{PL})}. \tag{2.7}$$

Since all the dimensions of the ring and pickup-loop are on the order of 1 mm, and we can measure voltages ratios to better than 5%, we can measure λ bigger than 1 mm.

The critical current density in the ring J^c can also be calculated in the weak stiffness limit. In this case we define a critical current in the coil I_{IC}^c, as the current at which the linearity between the signal ΔV_R^{max} and coil current I_{IC} breaks. This happens when $A_{IC}(R_{in})$ reaches a certain critical value $A_{IC}^c(R_{in})$. We take $A_{IC}^c(R_{in})$ from the numerical calculation presented in Fig. 2.6 using I_{IC}^c. The critical current density in the weak stiffness limit is

$$J_{weak}^c = \frac{c}{4\pi\lambda^2}A_{IC}^c(R_{in}). \tag{2.8}$$

Similarly, the coherence length is given by [1]

$$\xi_{weak} = \frac{\phi_0}{2\sqrt{3}\pi A_{IC}^c(R_{in})}. \tag{2.9}$$

Since $\lambda \sim 1$ mm, and the critical current is on the order of 1 mA, $[A_{IC}^c(R_{in}) \sim 10^{-2}$ Oe-cm] we can measure $J^c \sim 1$ Acm^{-2}, and $\xi \sim 1$ μm.

2.3.1.2 Strong Stiffness

In the strong stiffness case, the total vector potential experienced by the ring $\mathbf{A}_T$ is a sum of $\mathbf{A}_{IC}$ and, in the Coulomb gauge,

$$\mathbf{A}_R(\mathbf{r}) = \frac{1}{c} \int\limits_{Ring} \frac{\mathbf{J}(\mathbf{r}')d^3\mathbf{r}'}{|\mathbf{r} - \mathbf{r}'|}, \tag{2.10}$$

namely, $\mathbf{A}_T = \mathbf{A}_{IC} + \mathbf{A}_R$. Using the fact that $\nabla^2(1/r) = -4\pi\delta(\mathbf{r})$ and the London equation one finds that

$$\nabla^2 \mathbf{A}_R = \frac{1}{\lambda^2(r)} \left(\frac{\Phi_{IC}}{2\pi r}\hat{\varphi} + \mathbf{A}_R \right) \tag{2.11}$$

where λ is infinite outside of the ring and a constant inside of it. In cylindrical coordinates, $\mathbf{A}_R = A(z, r)\hat{\varphi}$, and with the coordinate transformation

$$\mathbf{r}/R_{PL} \to \mathbf{r}, \mathbf{A}_R/\mathbf{A}_{IC}(R_{PL}) \to A, \lambda/R_{PL} \to \lambda \tag{2.12}$$

the equation in the ring becomes

$$\frac{\partial^2 A}{\partial z^2} + \frac{\partial^2 A}{\partial r^2} + \frac{1}{r}\frac{\partial A}{\partial r} - \frac{A}{r^2} = \frac{1}{\lambda^2}\left(A + \frac{1}{r}\right) \tag{2.13}$$

but now r, z, and λ are in units of R_{PL}, and A is in units of $A_{IC}(R_{PL})$. Outside of the ring, the right hand side of the equation is zero. The solution of this equation, evaluated at R_{PL}, is the quantity one would measure with a single pickup-loop as indicated in Eq. 2.5.

We solved Eq. 2.13 for different λ values and our LSCO ring parameters with both the Comsol 5.2a and FreeFem++ softwares [9]. We used finite elements in a box $[-L_z, L_z] \times [0, L_r]$ where $L_z = L_r = 8$. Dirichlet boundary conditions are imposed at $z = \pm L_z$, $r = 0$, and $r = L_r$. Maximal mesh spacing is set to be $h = 0.01$ in the ring and its immediate vicinity, and $h = 0.25$ elsewhere. The total vector potential A_T for $\lambda/R_{PL} = 0.1/13$, and for all values of r and z in the ring's cross section is presented in Fig. 2.7. Clearly, the vector potential hence the current is strongest

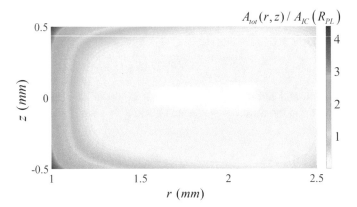

Fig. 2.7 A_T distribution inside the ring. The total vector potential obtained from the solution of Eq. 2.13 and the vector potential of the inner-coil A_{IC}, as a function of r and z for $\lambda/R_{PL} = 0.1/13$

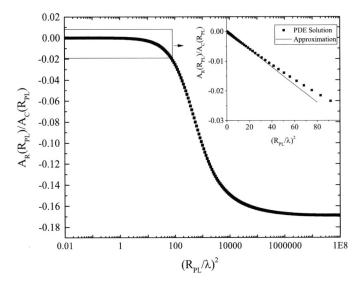

Fig. 2.8 Solution of the Stiffnessometer PDE. A semi-log plot of the solution of Eq. 2.13 evaluated at the pickup coil radius, for different values of $(R_{PL}/\lambda)^2$. The inset shows the behavior for large λ. The solid line is given by Eq. 2.7

close to the inner radius of the ring. They decay towards the center of the ring. The solution at $r = 1$ and $z = 0$ and a range of λ values is presented in Fig. 2.8 on a semi-log plot. The upper right inset is a zoom-in on the long λ region emphasized by a rectangle. The solid line represents Eq. 2.7 again with the LSCO ring parameters. There is a good agreement between the PDE solution at long λ and the weak-stiffness approximation.

2.3.2 Gradiometer

At this step we convert between the signal as detected by a gradiometer to the vector potential calculated above for a single pickup-loop. We find a conversion factor, the "G factor", from the vector potential evaluated at the position on a single pickup-loop A^{PL} to the differences in the vector potential generated by the gradiometer ΔA^G. This has to be done for both the ring and the inner-coil. The vector potential of a ring on the pickup-loop depends on its height z from the plane of the loop according to $A = 2\pi m R_{PL}^2 / (R_{PL}^2 + z^2)^{\frac{3}{2}}$. Therefore, for a ring and our gradiometer

$$
\frac{A_R^G(z)}{A_R^{PL}} = \frac{-2R_{PL}^3}{(R_{PL}^2 + (z + \Delta z_{PL})^2)^{\frac{3}{2}}} + \frac{4R_{PL}^3}{(R_{PL}^2 + z^2)^{\frac{3}{2}}} +
$$
$$
+ \frac{-2R_{PL}^3}{(R_{PL}^2 + (z - \Delta z_{PL})^2)^{\frac{3}{2}}},
\tag{2.14}
$$

where $\Delta z_{PL} = 7$ mm is the separation between the different groups of gradiometer windings. The difference between the maximum and minimum of this function, $\Delta A_R^G / A_R^{PL} = 1.7$, is the conversion factor for the ring.

Next we convert from A_{IC}^{PL} to ΔA_{IC}^G. For this purpose we plot by the green line in the inset of Fig. 2.6 the vector potential generated by our coil at R_{PL} as a function of z, $A_{IC}^{PL}(z)$. The function

$$
\frac{A_{IC}^G(z)}{A_{IC}^{PL}} = \frac{-2A_{IC}^{PL}(z + \Delta z_{PL}) + 4A_{IC}^{PL}(z) - 2A_{IC}^{PL}(z - \Delta z_{PL})}{A_{IC}^{PL}(0)}
\tag{2.15}
$$

is also plotted in the inset by the blue line. The difference between the maximum and minimum of this function is the conversion factor for the inner-coil. We find numerically that $\Delta A_{IC}^G / A_{IC}^{PL} = 0.47$. Thus

$$
\frac{\Delta V_R^{max}}{\Delta V_{IC}^{max}} = 3.62 \frac{A_R^{PL}}{A_{IC}^{PL}}
\tag{2.16}
$$

In Fig. 2.8 we see that when the penetration depth is very short, $A_R^{PL} / A_{IC}^{PL} = -0.17$. Multiplying the absolute value of this number by 3.62 we expect a saturation value of $\Delta V_R^{max} / \Delta V_C^{max} = 0.615$. The measured value, however, is 0.816 as seen in the inset of Fig. 2.3. To resolve this issue, we extract the G factor experimentally from the data by dividing the measured voltages ratios by the numerical saturation value. For the presented data of LSCO $x = 0.12$ this yields $G = 4.8$.

We now extract the penetration depth from the data in the inset of Fig. 2.3 using the conversion factor, and the PDE solution presented in Fig. 2.8. The extracted λ versus temperature is depicted in Fig. 2.9. Clearly we can determine λ as long as 2.5 ± 0.044 mm. The shortest λ we can pinpoint is 0.1 ± 0.04 mm. In order to determine the behavior near $T_c = 28$ K, $\lambda(T) \sim (1 - T/T_c)^{-\nu}$, we show a log-

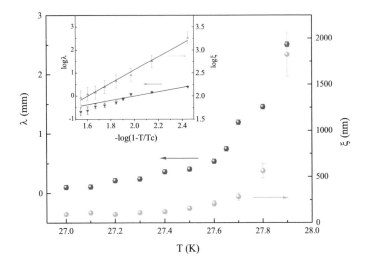

Fig. 2.9 London penetration depth and coherence length. λ (T) and ξ (T) are calculated as described in the text. Inset: log-log plot of λ and ξ as a function of $1 - T/T_c$

log plot of the data in the inset of Fig. 2.9. Linear fit yields a critical exponent $\nu = 1.0 \pm 0.1$, whereas mean field theory predicts $\nu = 0.5$ [1].

The calculation of the critical current needs an adjustment

$$J^c = \frac{1}{\mu_0 \lambda^2} A_T^{\max}(\lambda) = J^c_{weak} \left[\frac{A_T^{\max}(\lambda)}{A_{IC}(R_{PL})} \frac{R_{in}}{R_{PL}} \right] \qquad (2.17)$$

where $A_T^{max}(\lambda)$ stands for the maximum total vector potential in the ring evaluated for λ at the temperature at which the critical current is measured. Similarly, the coherence length is given by

$$\xi = \frac{\phi_0}{2\pi\sqrt{3}A_T^{\max}(\lambda)} = \frac{\xi_{weak}}{\frac{A_T^{\max}(\lambda)}{A_{IC}(R_{PL})}\frac{R_{in}}{R_{PL}}} \qquad (2.18)$$

For example, using this strong stiffness approach we find that at $T = 27.0$ K ($I_c = 0.75$ mA), $\lambda = 0.1$ mm, $J^c_{weak} = 50$ Acm^{-2} and $\xi_{weak} = 29.3$ nm, $A_T^{max}(\lambda)/A_{IC}(R_{PL}) = 4.5$, $J^c = 17.3$ Acm^{-2}, and $\xi = 83$ nm.

We present $\xi(T)$ in Fig. 2.9, while log-log plot is presented in the inset. We linearly fit the data and get critical exponent of 1.43 ± 0.06, whereas Mean Field theory predicts $\nu = 0.5$ [1].

2.4 Conclusions

We demonstrated that the Stiffnessometer can measure penetration depth two orders of magnitude longer, or stiffness four orders of magnitude smaller than ever before. This allows us to perform measurement closer to T_c and explore the nature of the superconducting phase transition, or determine the stiffness at low T in cases where it is naturally very weak. The Stiffnessometer also allows measurements of very small critical current or long coherence lengths, properties which again are useful close to T_c. The measurements are done in zero magnetic field with no leads, thus avoiding demagnetization, vortices, and out-of-equilibrium issues.

References

1. Tinkham M (2004) Introduction to superconductivity. Courier Corporation
2. Uemura YJ, Luke GM, Sternlieb BJ, Brewer JH, Carolan JF, Hardy WN, Kadono R, Kempton JR, Kiefl RF, Kreitzman SR et al (1989) Universal correlations between T_c and $\frac{n_s}{m^*}$ (carrier density over effective mass) in high-T_c cuprate superconductors. Phys Rev Lett 62:2317–2320
3. Homes CC, Dordevic SV, Strongin M, Bonn DA, Liang R, Hardy WN, Komiya S, Ando Y, Yu G, Kaneko N et al (2004) A universal scaling relation in high-temperature superconductors. Nature 430:539
4. Prozorov R, Giannetta RW (2006) Magnetic penetration depth in unconventional superconductors. Supercond Sci Technol 19:R41
5. Morenzoni E, Prokscha T, Suter A, Luetkens H (2004) Khasanov R (2004) Nano-scale thin film investigations with slow polarized muons. J Phys Condens Matter 16:S4583
6. Lamhot Y, Yagil A, Shapira N, Kasahara S, Watashige T, Shibauchi T, Matsuda Y, Auslaender OM (2015) Local characterization of superconductivity in $BaFe_2(As_{1-x}P_x)_2$. Phys Rev B 91:060504
7. Shay M, Keren A, Koren G, Kanigel A, Shafir O, Marcipar L, Nieuwenhuys G, Morenzoni E, Suter A, Prokscha T, Dubman M, Podolsky D (2009) Interaction between the magnetic and superconducting order parameters in a $La_{1.94}Sr_{0.06}CuO_4$ wire studied via muon spin rotation. Phys Rev B 80:144511
8. Talantsev EF, Tallon JL (2015) Universal self-field critical current for thin-film superconductors. Nature Commun 6:7820
9. Hecht F (2012) New development in FreeFem++. J Numer Math 20(3–4):251–265. 65Y15

Chapter 3
The Nature of the Phase Transition in the Cuprates as Revealed by the Stiffnessometer

The Stiffnessometer, as explained in Chap. 2, gives an opportunity to measure stiffness close to Tc, and to gain new insight with respect to the superconducting phase transition. In this chapter, we utilize this new tool to examine the phase transition in LSCO, and the possibility for bulk 2D SC. We start by recapitulating the Stiffnessometer method, but here we generalize the London equation and represent the stiffness as a tensor. This, as will be shown, will allow to measure both inplane and out ot plane stiffness. We will then show and discuss the results, and compare them to measurements of Low Energy μSR.

3.1 Stiffnessometer

The Stiffnessometer is based on the fact that outside an infinitely long coil, the magnetic field is zero while the vector potential $\mathbf{A}$ is finite. When such a coil is threaded through a superconducting ring, the vector potential leads to supercurrent density $\mathbf{J}$ according to the London equation $\mathbf{J} = -\overline{\rho}_s \mathbf{A}$, where $\overline{\rho}_s$ is the stiffness tensor. This current flows around the ring and generates a magnetic moment. We detect this moment by moving the ring and the inner-coil (IC) rigidly relative to a Gradiometer, which is a set of pickup loops wound clockwise and anticlockwise. The Gradiometer is placed in the center of a bigger coil which is used to cancel stray field on the sample. The experimental set-up, our coil and ring are presented in Fig. 3.1a. The voltage generated in the Gradiometer by the inner coil and the sample movement is measured by a SQUID magnetometer. The measurements are done in zero gauge-field cooling procedure, namely, the ring is cooled to a temperature below T_c, and only then the current in the inner coil is turned on. It is the change in magnetic flux inside the inner coil which creates an electric field in the ring, and sets persistent currents in motion.

This chapter was published on 2019 in Nature Communications 10, Article number: 2463, https://doi.org/10.1038/s41467-019-10480-x

I. Kapon, *Searching for 2D Superconductivity in La2−x Srx CuO4 Single Crystals*,
Springer Theses, https://doi.org/10.1007/978-3-030-23061-6_3

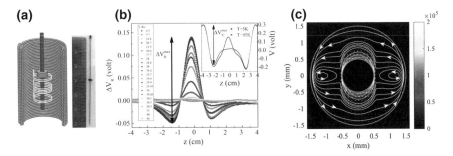

Fig. 3.1 Stiffnessometer **a** An illustration of the Stiffnessometer operation principal and a photo of typical ring and coil with 2400 windings. A long coil is threaded through a ring and they both move with respect to a Gradiometer which is connected to a SQUID. The SQUID measures the flux through the Gradiometer and hence the average vector potential on it $\langle A^{\theta} \rangle$. **b** Temperature dependence of an LSCO $x = 0.125$ c-ring signal as measured by the Stiffnessometer with $I = 1$ mA in the inner-coil. The data presented are after subtraction of the coil contribution, $\Delta V_R(z)$, as explained in the text and in Ref. [1]. The inset shows raw Stiffnessometer data for a temperature above and below T_c. The difference is due to the ring contribution. **c** The currents streamlines in the ring at midheight ($z = 0$) derived from the solution of Eq. 3.2 for the a-ring with $\lambda_c = 145$ μm and $\lambda_{ab} = 13.9$ μm. The false colors show the current intensity. Naturally the flow is not isotropic. Vortices develop on both sides of the x axis

To examine the orientation dependent response of LSCO to different directions of **A**, we cut two types of rings from a single crystal rod: "c-ring" where the crystallographic $\hat{c}$ direction is parallel to the ring symmetry axis, i.e. the supercurrent flows in the CuO_2 planes, and "a-ring" where the crystallographic $\hat{a}$ direction is parallel to the ring symmetry axis, i.e. the supercurrent travels both in the planes and between them. The rings, shown in Fig. 3.2a, have inner radius of 0.5 mm, outer radius of 1.5 mm and 1 height.

The inset of Fig. 3.1b presents raw Stiffnessometer data of c-ring taken with inner coil current of 1 mA. The vertical axis is the measured voltage by the SQUID. The horizontal axis is the position z of the ring relative to the center of the Gradiometer. The red data points are measured above T_c and represent the signal generated by the inner coil alone. The blue points are measured below T_c and correspond to the inner coil and the ring. The difference between them, $\Delta V_R(z)$, is the signal from the ring itself. This signal is shown in Fig. 3.1b for different temperatures. Between 4.5 and 27 K there is hardly any change in the signal, because the Stiffnessometer is not sensitive to short penetration depth compared to the sample size. However, above 28 K the signal drops dramatically fast with increasing temperature.

We define the peak-to-peak voltage of the rings and the inner coil, ΔV_R^{max} and ΔV_{IC}^{max} respectively, as shown in Fig. 3.1b. Their ratio holds the information about the stiffness, as we explain shortly. Figure 3.2a presents ΔV_R^{max} of both rings. These voltages are normalized by their maximal value for comparison purposes. We detect two different stiffness transition temperatures, $T_s^c = 30.1$ K for the c-ring, and a lower one $T_s^a = 29.4$ K for the a-ring. We also examine the influence of the inner coil current on the transition. Data corresponding to three different currents are shown

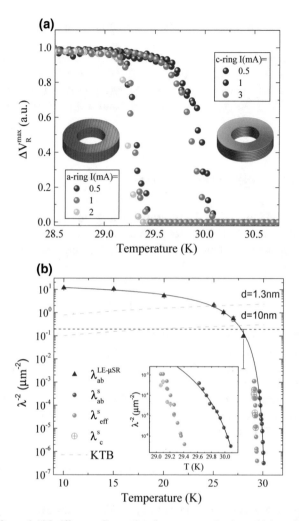

Fig. 3.2 LSCO $x = 0.125$ stiffness. **a** Comparison between a- and c-ring, which are demonstrated in the figure, as measured by the Stiffnessometer. The signal is normalized by the maximum measured ring voltage. Different transition temperatures are observed for the two kind of rings with 0.7 K difference between them. The transition does not depend on the applied current in the inner coil up to 1 mA. **b** Semi-log plot of λ_{ab}^{-2} as measured by LE-μSR (purple solid triangles) and Stiffnessometer (blue solid spheres). Black dashed line represents the sensitivity limit of LE-μSR. Black solid line is a fit to a phenomenological function described in the text. Dashed blue lines represent the KTB line for layer widths $d = 1.3$ nm and $d = 10$ nm. Green solid spheres represent the penetration depth of an a-ring from the Stiffnessometer, analyzed as if the ring is isotropic with λ_{eff} which is some combination of λ_{ab} and λ_c. Orange open symbols show λ_c obtained at the temperature range where their ratio is manageable numerically for analysis. The inset is a zoom in on temperatures close to the transitions

in the figure. Below 1 mA there is no change in the transition, which otherwise widens and appears at slightly lower temperature.

The Stiffnessometer data reveal a new phenomenon. There is a temperature range with finite 2D stiffness in the planes, although supercurrent cannot flow between them. In other words, upon cooling, the SC phase transition starts by establishing a global 2D stiffness, and only at lower temperature a true 3D superconductivity is formed.

3.1.1   Stiffnessometer Data Analysis

To analyze the data, we relate the measured voltage to the vector potential. Since SQUID measures flux, and the vector potential on the Gradiometer is proportional to the flux threading it, the ratio of the peak-to-peak voltages satisfies

$$\frac{\Delta V_R^{max}}{\Delta V_{IC}^{max}} = G \frac{\langle A_R^\theta(R_{PL})\rangle}{A_{IC}^\theta(R_{PL})} \tag{3.1}$$

where A_R^θ and A_{IC}^θ are the rings and inner coil vector potential components in the azimuthal direction $\hat{}$ respectively, R_{PL} is the Gradiometer radius, $\langle\rangle$ stands for averaging over the pickup loops, and G is a geometrical factor determined experimentally (see Supplementary Information).

In order to extract $\overline{\rho}_s$ from the voltages ratio of Eq. 3.1 we must determine the dependence of $\mathbf{A}_R(R_{PL})$ on the stiffness. This is done by numerically solving the combined Maxwell's and London's equation

$$\nabla \times \nabla \times \mathbf{A}_R = \overline{\rho}_s \left(\mathbf{A}_R + \frac{\Phi_{IC}}{2\pi r}\hat{\theta}\right) \tag{3.2}$$

where Φ_{IC} is the flux through the inner-coil, and $\overline{\rho}_s$ is finite only inside the ring. For c-ring $\overline{\rho}_s$ is merely a scalar and equals λ_{ab}^{-2}. For a-ring, it is diagonal in Cartesian coordinates, with $\rho_{xx} = \lambda_c^{-2}$ and $\rho_{yy} = \rho_{zz} = \lambda_{ab}^{-2}$.

We solve Eq. 3.2 numerically for our rings geometry and various λ_{ab} and λ_c with FreeFEM++ [2] and Comsol 5.3a. The c-ring solution, which is sensitive to λ_{ab} only, is discussed in Chap. 2. Using Eq. 3.1, the numerical solution, and the data in Fig. 3.1b we extract $1/\lambda_{ab}^2$, and plot it in Fig. 3.2b on a semi-log scale (blue solid spheres).

In order to extract λ_c we have to know λ_{ab} at the temperatures of interest. As can be seen from Fig. 3.2a the c-ring Stiffnessometer measurements are in saturation just when a-ring stiffness becomes relevant. Therefore, we applied LE-μSR to the same samples.

3.2 LE-μSR

In LE-μSR spin polarized muons are injected into a sample. By controlling the muons energy E between 3–25 keV, the muons stop with high probability at some chosen depth inside the sample while keeping their polarization intact. When an external magnetic field is applied, the muon spin rotates at the Larmor frequency corresponding to the field. Since the magnetic field decays in the sample on a length scale determined by λ, the frequency becomes smaller as the muons stop deeper in the sample.

For our LE-μSR measurements, the sample is a mosaic of plates cut in the ac crystallographic plane from the same LSCO $x = 0.125$ crystal used for the Stiff-nessometer. Each plate was mechanically polished to a roughness of few tens of nanometers. The plates were glued to a Nickel coated plate using silver paste (see Fig. 3.6 inset). We cooled the sample to 5 K in zero magnetic field. Then a transverse magnetic field was applied along the $\hat{\mathbf{a}}$ or $\hat{\mathbf{c}}$ directions, and we warmed to the desired temperature.

There are two methods by which one can extract the penetration depth. The simple method is to fit each data set (at each temperature and energy) to $A(t) = A_0 \exp(-t/T_2) \cos(\omega t)$. From this fit one can extract asymmetry, relaxation, and the average internal field as a function of average implantation depth and temperature.

Figure 3.3 summarizes the internal magnetic field as a function of implantation energy for different temperatures and field orientations. The field here is calculated by $B = \omega/2\pi\gamma$, where ω is the angular frequency of the muon polarization and γ is the gyromagnetic ratio. Noticeably, close to the surface and at low T, the magnetic field does not change with increasing implantation depth for $\mathbf{H} \parallel \mathbf{c}$. Only for energies above 5 keV does a linear trend of decay appears. This $10-20$ nm of "dead layer" could be a byproduct of the polishing process.

Fig. 3.3 Magnetic field as a function of implantation energy. Closed symbols are $\mathbf{H} \parallel \mathbf{c}$ and open symbols are $\mathbf{H} \parallel \mathbf{a}$. Straight lines are guides to the eye. The magnetic field below $E = 5$ keV does not fit the linear trend of the field decay, indicating a dead layer of about $10-20$ nm, possibly caused by the polishing treatment

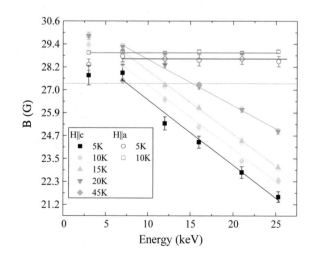

Fig. 3.4 Temperature
dependance of LE-μSR
parameters. The
measurement was done at
constant energy of 24 keV.
The magnetic field (**a**)
displays peculiar behavior
near T_c. Its magnitude below
T_c is larger than that of the
normal state. The asymmetry
(**b**) is constant until 20K,
where it starts to drop due to
magnetic freezing. The
magnetism is also exhibited
in an uprise of the decay rate
(**c**) at low temperatures

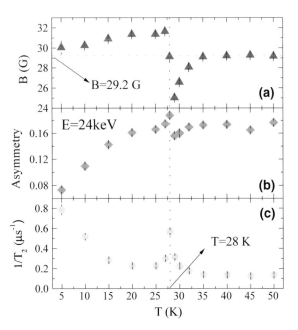

Figure 3.4 depicts the temperature dependence of the individual fit parameters
for the highest implantation energy. The magnetic field (panel (a)) seems to behave
erratically close to the phase transition into the superconducting state. We attribute
this behavior to demagnetization factor and mutual coupling between different pieces
of the sample. The asymmetry (panel (b)) decreases upon cooling since LSCO x
= 0.125 is known to have a magnetic phase concomitant with the superconducting
one [3–5]. The muon spin relaxation (panel (c)) has a peak at the critical temperature,
which is also unusual.

The presence of magnetism could be detrimental to our analysis if it depends on
depth. To verify that this is not the case, we perform zero field (ZF) measurements
for different implantation energies at $T = 5$ K well below T_c and for $T = 30$ K
above T_c. The results are presented in Fig. 3.5. Fast relaxation and reduction of the
asymmetry are observed at low temperature due to local random fields originating
from the magnetic stripes in the sample. Nevertheless, there is no change in the
magnetic relaxation with implantation depth.

The more sophisticated analysis method, which we use rather than the simple
one, takes into account the stopping depth probability of the muons. The stopping
profile $p(x, E)$, where x is stopping depth, is simulated by the TRIM.SP Monte Carlo
code [6]. Figure 3.6 presents the LSCO stopping profiles for different implantation
energies. For each energy, we fit the function

$$p(x, E) = \frac{p_0(x_0 - x)^3}{\exp[(x_0 - x)/\xi] - 1} H(x_0 - x). \tag{3.3}$$

Fig. 3.5 Depth independent magnetism in LSCO x = 0.125. Asymmetry versus time at $T = 5$ K (close symbols) and $T = 30$ K (open symbols) for different implantation energies. The signal does not change as a function of energy at low temperatures, justifying a depth independent relaxation component (see main text)

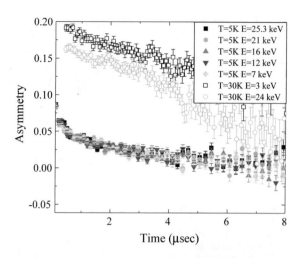

Fig. 3.6 Muon stopping profiles. The probability distribution of a muon to stop at some depth x inside the sample for different implantation energies. The inset shows the LSCO x = 0.125 single crystal samples used in the experiment. All the pieces were polished to roughness of several nanometers. The crystallographic axes a and c are in the plane of the samples, and shown in the picture

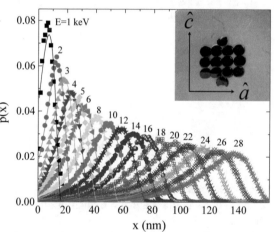

to this profile. Here x_0 is some cut-off position the muon cannot cross and is energy dependent, $H(x_0 - x)$ is Heaviside's function, and ξ and p_0 are energy dependent free parameters. The energy dependence of the fit parameters is

$$p_0(E) = \exp\left[-6.4 - 0.8\ln E - 0.18(\ln E)^2\right]$$
$$x_0(E) = 12 + 6E - 0.11E^2 + 0.0028E^3$$
$$\xi(E) = 2.77 + 0.49E - 0.0165E^2 + 0.0003E^3$$

We assume an exponential decay of the magnetic field along the direction perpendicular to the sample surface, x, resulting from the Meissner effect. In this case, the asymmetry is given by

Fig. 3.7 LE-μSR spectra.
Asymmetry as a function of
time for different muon
implantation energies for: **a**
H ‖ $\hat{c}$, $H = 26.7$ Oe,
$T = 20$ K, **b H** ‖ $\hat{c}$,
$H = 26.7$ Oe, $T = 30$ K, **c**
H ‖ $\hat{a}$, $H = 26.3$ Oe,
$T = 11$ K. A clear frequency
shift as a function of
implantation energy is
observed in (**a**). In the (**b**)
conditions, the
Stiffnessometer clearly
detects stiffness in the ab
plane (Fig. 3.2a), while no
frequency shift is observed
by LE-μSR within our
sensitivity. For **H** ‖ $\hat{a}$ (**c**)
there is no frequency shift at
all temperatures

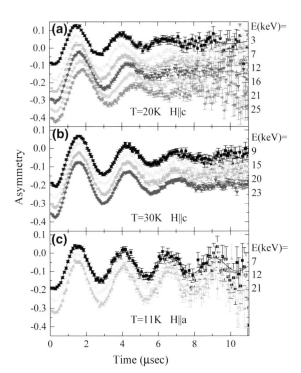

$$A(E, t) = A_0 e^{-t/u} \int_0^\infty p(x, E) \cos\left(\gamma B_0 e^{-x/\lambda} t\right) dx, \qquad (3.4)$$

where $1/u$ represents contributions to the relaxation from depth independent processes and B_0 is the magnetic induction outside of the sample and parallel to its surface.

Figure 3.7 presents asymmetry data for both magnetic field orientations and different implantation energies. Panels (a) and (b) show data for **H** ‖ $\hat{c}$ at two different temperatures, and panel (c) depicts data for **H** ‖ $\hat{a}$. The data sets are shifted vertically for clarity. We limit the presentation to temperatures above 10 K, since below it strong relaxation due to spin density wave order obscures the oscillatory signal.

For each temperature, we fit all data sets with energy larger than 5 keV due to the presence of a dead layer, using Eq. 3.4. In the fit A_0 is a free parameter, and λ, u and B_0 are shared. A_0 is free because the number of muons actually penetrating the sample varies with energy. u represents relaxation processes that are implantation depth independent such as magnetism or field variations perpendicular to x. These are taken into account as some Lorentzian probability distribution of the total internal magnetic field with FWHM of $2/u$. λ and B_0 are naturally common to each temperature.

At $T = 20$ K **H** ‖ $\hat{c}$, we observe a clear frequency shift as a function of implantation energy, indicating a Meissner state. However, for $T = 30$ K, where the Stiffnessometer clearly shows $\rho_{ab} > 0$, we could not detect any change in frequency, even though we used high statistics data acquisition of 24 million muons for $E = 23$keV and 8 million for the rest. This can be explained from the fact that the penetration depth here is much longer than the muon stopping length scale of the order of hundred nanometer. When $\mathbf{H} \parallel \hat{\mathbf{a}}$ we did not observe any frequency shift at all temperatures, even though the sample is in the Meissner state.

We fit Eq. 3.4 to our LE-μSR data and extract λ_{ab}. We add the results to Fig. 3.2b. There is a gap between the available data from the two techniques because the longest penetration depth that LE-μSR can measure, represented by the horizontal dashed line in the figure, is much smaller than the shortest λ for which the Stiffnessometer is sensitive to. The function $\lambda_{ab}^{-2} = C_0 \exp\left[C_1 / \left(1 + C_3 \left(1 - T/T_c\right)^\delta\right)\right]$ is fitted to the combined data and serves for interpolation. Since at $T = 10$ K we could only measure λ_{ab} and not λ_c, we deduce an anisotropy $\lambda_c(0)/\lambda_{ab}(0) \geq 10$, as was observed in μSR, optical, and surface impedance measurements [7–9].

3.3 Results

We are now in position to extract λ_c from Eq. 3.1, Eq. 3.2 and the Stiffnessometer a-ring data in Fig. 3.2a. In this case, two coupled partial differential equations must be solved, where λ_{ab} is determined from the c-ring interpolation.

The gauge choices are as follows: Inside the ring, applying divergence to Eq. 3.2 yields the gauge $\nabla \cdot \left(\overline{\rho}_s \mathbf{A}_{tot}\right) = 0$, where $\mathbf{A}_{tot} = \mathbf{A_R} + \mathbf{A_{IC}}$. This gauge also enforces the continuity equation for the current density $\mathbf{J} = \overline{\rho}_s \mathbf{A}_{tot}$. Outside the ring we apply the Coulomb gauge $\nabla \cdot \mathbf{A}_{tot} = 0$, which is also used to determine $\mathbf{A_{IC}}$ and $\mathbf{A}_{tot}$ in the isotropic case. The boundary conditions are $A(\infty) = 0$. In practice, infinity is understood as the domain surface, and the domain is taken to be large enough so that finite-domain effects are negligible. The domain of the problem is defined as a cylinder with height 100 times that of the ring, i.e. $7.7R_{PL}$ and outer radius 100 times that of the ring, i.e. $11.5R_{PL}$. Since no current can cross the ring surface, we demand $\mathbf{J}_\perp(\mathbf{r}_{in}) = \mathbf{J}_\perp(\mathbf{r}_{out}) = 0$ where $\perp$ stands for the direction perpendicular to the surface, and r_{in} (r_{out}) is the inner (outer) radius of the ring. Finally, from the absence of a surface field, we demand $\Delta\mathbf{A}_\parallel(\mathbf{r}_{in}) = \Delta\mathbf{A}_\parallel(\mathbf{r}_{out}) = 0$, where $\Delta\mathbf{A}_\parallel$ stands for the difference between the vector potential parallel to the surface inside the ring and outside of it.

Figure 3.8 shows the numerical results of the vector potentials ratio that appears in Eq. 3.1 as a function of (a) $(R/\lambda_{ab})^2$ and (b) $(R/\lambda_c)^2$ for $\lambda_{ab} = 13.9$ μm at $T = 29.16$ K. In our analysis, λ_{ab} is extracted from the c-ring data in the isotropic case. Then, for each temperature, the corresponding λ_{ab} is used to generate the result in panel (b), and combining with the a-ring data λ_c is extracted.

Figure 3.9 presents the numeric solution of the ring vector potential A_R at $z = 0$ plane (midheight of the ring), calculated for LSCO x = 0.125 a-ring at $T = 29.16$ K with $\lambda_{ab} = 13.9$ μm, extracted from the extrapolation function presented in the main text, and $\lambda_c = 145$ μm. Panel (a) shows the azimuthal part of $\mathbf{A}$, whereas panel (b) the radial one.

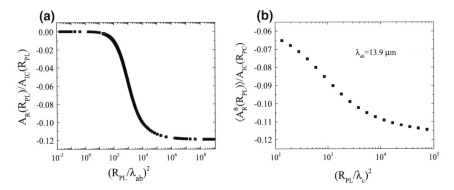

Fig. 3.8 Extracting the stiffness. Numerical results of the vector potentials ratio as appears in Eq. 3.1 as a function of **a** $(R/\lambda_{ab})^2$ and **b** $(R/\lambda_c)^2$ for $\lambda_{ab} = 13.9$ μm at $T = 29.16$ K

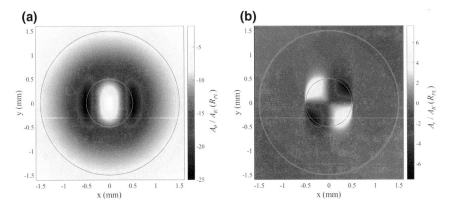

Fig. 3.9 Vector potential for LSCO x = 0.125 a-ring. Numeric solution of the radial (**a**) and azimuthal (**b**) components of the vector potential inside the ring at z = 0 for $\lambda_{ab} = 13.9$ μm and $\lambda_c = 145$ μm

Figure 3.10 shows the absolute value of the current density **J** inside the rings for two cuts at fixed angles: (a) xz plane, (b) yz plane. At the xz plane, the current concentrates at a very thin layer close to the ring inner rim, while in the yz plane the current penetrates further into the bulk. This corresponds, of course, to the large difference in the penetration depth in the two directions.

Finally, Fig. 3.11 shows the magnetic field generated by the ring as calculated from the curl of $\mathbf{A_R}$. The penetration pattern of the field is of an ellipse due to the penetration depths anisotropy.

Currently, we manage to extract λ_c for only few temperatures close to T_s^a, where the anisotropy ratio is not too big and numerically solvable. These values of λ_c are presented as orange open symbols in Fig. 3.2b. The SC currents in the ring at z=0 emerging from the numerical solution for $T = 29.16$ K are depicted in Fig. 3.1c by combined contour and quiver plots.

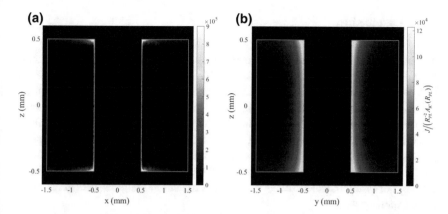

Fig. 3.10 Current density simulation inside LSCO x = 0.125 a-ring. False color map of the current density distribution in a ring with $\lambda_{ab} = 13.9\,\mu$m and $\lambda_c = 145\,\mu$m in the (**a**) xz plane and (**b**) yz plane. Most of the current concentrates on the inner rim

Fig. 3.11 Magnetic field False color map of the magnetic field z component inside the a-ring and its vicinity for $\lambda_{ab} = 13.9\,\mu$m and $\lambda_c = 145\,\mu$m

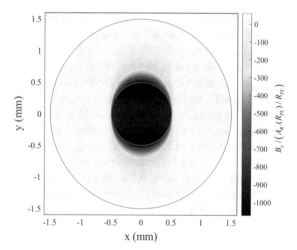

For all a-ring Stiffnessometer data we also applied the c-ring stiffness extraction method ignoring the anisotropy. By doing so we determine an effective stiffness λ_{eff}^{-2}, which is some combination of λ_{ab}^{-2} and λ_c^{-2}. These values are presented as green solid spheres in Fig. 3.2b. λ_{eff}^{-2} is larger than λ_c^{-2} but shows the same trend and indicates two transition temperatures.

3.4   Discussion

The observation of two transition temperatures is awkward; a material should have only one SC critical temperature. One possible speculation for this result is a finite size effect, namely, if the rings could be made bigger the difference between the two transition temperatures would diminish. This, however, cannot be the case since the sample size is taken into account when extracting the stiffness. Bigger samples should lead to the same λ values. A more plausible explanation is that the phase transition starts in the form of wide superconducting filaments [10] or finite width sheets [11] in the planes, but disconnected in the third direction. Whether this is the case, or our result indicates a new type of phase transitions, requires further and more local experiments.

The two transition temperatures suggest that there is a temperature range in which the system behaves purely as 2D. Therefore, we examine whether λ_{ab}^{-2} follows the KTB behavior. At the KTB transition, the stiffness should undergo a sharp increase (a "jump") at a temperature T_{KTB} that satisfies $\lambda^{-2} = \gamma T_{KTB}$, where $\gamma = \frac{8k_B e^2 \mu_0}{\pi \hbar^2 d}$ and d is the layer thickness [12]. We plot the line $\lambda^{-2} = \gamma T$ in Fig. 3.2b for thickness $d = 1.3$ nm of one unit cell (u.c.) and for $d = 10$ nm of about 8 u.c., both in cyan dashed lines. Clearly, the KTB line for thickness of one u.c. does not intersect λ_{ab}^{-2} where it exhibits a jump. The line for $d = 10$ nm, however, does seem to intersect at the beginning of a jump. Thus, for the transition to be of the KTB nature, an effective layer of about 8 unit cells and more is needed.

In summary, using new magnetic-field-free superconducting stiffness tensor measurements, which are sensitive to unprecedented long penetration depths, on the order of millimeters, and which are not affected by demagnetization factors or vortices, we shed new light on the SC phase transition in LSCO $x = 1/8$. In this compound, there is a temperature interval of 0.7 K where SC current can flow in the CuO_2 planes but not between them. When stiffness develops in both directions, the ratio of penetration depths obeys $\lambda_c/\lambda_{ab} \geq 10$.

3.5   Supplementary Information

Materials

The LSCO single crystals were grown using Traveling Solvent Floating Zone furnace, annealed in Argon environment at $T = 850$ C for 120 h to release internal stress, and oriented by Laue x-ray diffraction. Stiffnessometer samples were cut into a shape of rings using pulsed Laser ablation, after which the rings were annealed again. LE-μSR samples were mechanically polished using diamond paste. They were treated eventually with 20 nm alumina suspension. The resulting roughness of few tens of nanometers was determined by Atomic Force Microscope (AFM). A typical AFM data is presented in Fig. 3.12.

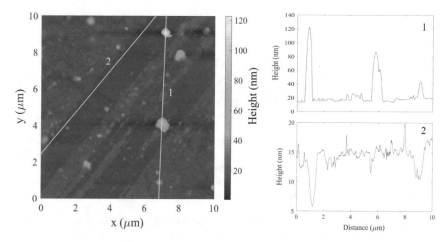

Fig. 3.12 Sample surface roughness. AFM image of one polished LSCO x = 0.125 plate treated with 20 nm Alumina suspension. Height profiles along two lines are presented, demonstrating fairly smooth surface

Stiffnessometer

The Stiffnessometer is an add-on to a Cryogenic SQUID magnetometer. The components of the experiment shown in Fig. 3.1 in the main text are as follows: The inner coil is 60 mm long with a 0.05 mm diameter wire and two layers of windings. It is wound on top of a 0.54 mm diameter polyamide tube. The outer diameter of the coil is 0.74 mm, and it has 40 turns per millimeter. The second order Gradiometer is 14 mm high, with inner diameter 25.9 mm, outer diameter 26.3 mm, and made from 0.2 mm diameter wire. We take $R_{PL} = 13 \pm 0.15$ mm. The Gradiometer is constructed from three groups of windings distanced 7 mm apart from each other. The upper and lower ones have two loops wound clockwise, while the center windings have four loops wound anticlockwise. Numeric evaluation of the G factor in Eq. 3.1 using the Gradiometer dimensions gives a reasonable result. However, the G factor used here is extracted experimentally. As shown in Fig. 3.2 in the main text, the signal from the rings ΔV_R^{max} saturates at $T \ll T_c$. It happens when the penetration depth is much smaller than the ring dimensions. The ratio between the voltages saturation value to the vector potentials ratio calculated numerically gives G.

References

1. Kapon I, Golubkov K, Gavish N, Keren A (2017) Stiffnessometer, a magnetic-field-free superconducting stiffness meter and its application. arXiv:1705.00624
2. Hecht F (2012) New development in freefem++. J Numer Math 20:251–265
3. Panagopoulos C, Tallon JL, Rainford BD, Xiang T, Cooper JR, Scott CA (2002) Evidence for a generic quantum transition in high-T_c cuprates. Phys Rev B 66:064501. https://link.aps.org/doi/10.1103/PhysRevB.66.064501

4. Suzuki T, Goto T, Chiba K, Shinoda T, Fukase T, Kimura H, Yamada K, Ohashi M, Yamaguchi Y (1998) Observation of modulated magnetic long-range order in $La_{1.88}Sr_{0.12}CuO_4$. Phys Rev B 57:R3229. https://link.aps.org/doi/10.1103/PhysRevB.57.R3229

5. Kimura H, Hirota K, Matsushita H, Yamada K, Endoh Y, Lee S-H, Majkrzak CF, Erwin R, Shirane G, Greven M, Lee YS, Kastner MA, Birgeneau RJ (1999) Neutron-scattering study of static antiferromagnetic correlations in $La_{2-x}Sr_xCu_{1-y}Zn_yO_4$. Phys Rev B 59:6517–6523. https://link.aps.org/doi/10.1103/PhysRevB.59.6517

6. Eckstein W (2013) Computer simulation of ion-solid interactions, vol. 10. Springer Science & Business Media

7. Homes CC, Dordevic SV, Strongin M, Bonn DA, Liang R, Hardy WN, Komiya S, Ando Y, Yu G, Kaneko N et al (2004) A universal scaling relation in high-temperature superconductors. Nature 430:539. https://doi.org/10.1038/nature02673

8. Shibauchi T, Kitano H, Uchinokura K, Maeda A, Kimura T, Kishio K (1984) Anisotropic penetration depth in $La_{2-x}Sr_xCuO_4$. Phys Rev Lett 72:2263–2266. https://link.aps.org/doi/10.1103/PhysRevLett.72.2263

9. Dordevic SV, Komiya S, Ando Y, Wang YJ, Basov DN (2005) Josephson vortex state across the phase diagram of $La_{2-x}Sr_xCuO_4$: a magneto-optics study. Phys Rev B **71**, 054503. https://link.aps.org/doi/10.1103/PhysRevB.71.054503

10. Davis SI, Ullah RR, Adamo C, Watson CA, Kirtley JR, Beasley MR, Kivelson SA, Moler KA (2018) Spatially modulated susceptibility in thin film $La_{2-x}Ba_xCuO_4$. Phys Rev B 98:014506. https://link.aps.org/doi/10.1103/PhysRevB.98.014506

11. Pekker D, Refael G, Demler E (2010) Finding the elusive sliding phase in the superfluid-normal phase transition smeared by c-axis disorder. Phys Rev Lett 105:085302. https://link.aps.org/doi/10.1103/PhysRevLett.105.085302

12. Nelson DR, Kosterlitz JM (1977) Universal jump in the superfluid density of two-dimensional superfluids. Phys Rev Lett 39:1201. https://link.aps.org/doi/10.1103/PhysRevLett.39.1201

Chapter 4
Opening a Nodal Gap by Fluctuating Spin-Density-Wave in Lightly Doped La$_{2-x}$Sr$_x$CuO$_4$

This chapter describes a different work we performed to investigate whether the spin or charge degrees of freedom were responsible for the nodal gap in underdoped cuprates. It is based on our paper published in Physical Review B journal [1].

4.1 Introduction

There are several indications by now from angle-resolved photoemission spectroscopy (ARPES) that in underdoped cuprates a gap opens at the Fermi surface in the diagonal (nodal) direction [2–6]. In La$_{2-x}$Sr$_x$CuO$_4$ (LSCO) this nodal gap (NG) extends to $x = 8\%$. At doping around $x = 12.5\%$ samples develop a charge-density-wave (CDW) below $T \approx 100$ K [7]. Traces of antiferromagnetism (AFM) in the form of spin-density-waves (SDW) [8] or spin-glass [9] appear at doping up to $x = 12.5\%$ and temperatures $T \approx 10$ K. It is therefore natural to speculate that one of these symmetry breaking phenomena is responsible for the opening of a nodal gap. In this work, we would like to clarify which one is the most likely. Our strategy is to carefully examine a sample which is known to have, at least, both AFM and SDW order, and opens a nodal gap at low temperatures. The sample is LSCO with $x = 1.92\%$ [2].

Previous neutron diffraction measurements on LSCO $x = 1.92\%$ [2] showed a magnetic Bragg peak at the AFM wave vector $\mathbf{Q}_{AF}$ below $T = 140$ K, and two satellites that stand for static SDW order (on top of the AFM one). The satellites appear below $T = 30$ K. Like in Matsuda et al. [8], there are two domains in the sample. We focus on one of them, in which the AFM peak is observed when scanning near (1, 0, 0), with no contribution from SDW. In contrast, the SDW peaks are observed when scanning near (0, 1, 0), with no contribution from the AFM peak. Neutron scattering detects the component of spin fluctuations perpendicular to the

© Springer Nature Switzerland AG 2019
I. Kapon, *Searching for 2D Superconductivity in La$_{2-x}$Sr$_x$CuO$_4$ Single Crystals*,
Springer Theses, https://doi.org/10.1007/978-3-030-23061-6_4

momentum transfer $\mathbf{q}$ [10]. Hence, the SDW fluctuations are perpendicular to the AFM order. ARPES measurements on the same sample found that a nodal gap opens below $T_{NG} = 45$ K [2]. Even though there is a temperature mismatch between the NG and SDW appearance, the two phenomena might be related. Moreover, CDW in LSCO $x = 1.92\%$ is expected to be very weak [11], and indeed this sample is out of the CDW dome [7, 12]. Therefore, a priori, CDW is not expected to generate the nodal gap.

Here we add to the available ARPES and neutron diffraction data, inelastic neutron scattering (INS) and X-ray diffraction data on the same piece of LSCO $x = 1.92\%$. We show that the fluctuating SDW amplitude of the frequency where it is the strongest, decreases at a temperature equal to T_{NG} within experimental error. In addition, we could not find any indications for CDW in our sample. We argue that these findings explain the previously measured 15 K discrepancy between the SDW freezing and the opening of a NG, and tie the latter to fluctuating SDW.

4.2   Results

The neutron experiment was performed at Rita-II, the cold neutrons triple axis spectrometer at the Paul Scherrer Institut. Throughout this paper, we work in orthorhombic notation, with cell parameters $a = 5.344$ Å, $b = 5.421$ Å and $c = 13.14$ Å at $T = 2$ K. In this notation, the tetragonal 2D $\mathbf{Q}_{AF} = (1/2, 1/2, 0)$ is equivalent to $(0, 1, 0)$ in reciprocal lattice units (r.l.u) of $2\pi/a$. More information is available in the Methods section. In Fig. 4.1 we present a false color map of neutron counts versus energy transfer $\hbar\omega$ and momentum transfer $\mathbf{q}$. The raw data, in this figure alone, is interpolated for presentation purpose. Data is presented at two temperatures, 2 and 50 K, which are below and above the freezing temperature of the incommensurate magnetic order of 30 K [2]. In both cases, strong intensity is observed at $\hbar\omega = 0$. This is due to high order contamination of the incoming beam scattering from a nuclear Bragg peak at $(0, 2, 0)$, despite the use of Br filter. Around $(0, 1, 0)$, the intensity extends to energy transfers as high as 8 meV for both temperatures, in a cone shape, which is in fact a poorly-resolved bottom part of an hourglass. This will be demonstrated subsequently. The scattering intensity is stronger at elevated temperatures. Interestingly, at $T = 2$ K spectral weight is missing at low energies, suggesting the presence of a soft gap for spin excitations. A similar spectrum, including the gap, was observed at the fully developed hourglass dispersion of $La_{1.875}Ba_{0.125}CuO_4$ [13], $La_{1.88}Sr_{0.12}CuO_4$ [14, 15], and $La_{1.6}Sr_{0.4}CoO_4$ [16], and also at La_2CuO_4 [17].

q-scans at specific constant energies at $T = 2$ K are presented in Fig. 4.2, showing the evolution of the SDW peaks with energy transfer. The intensities are shifted vertically for clarity. At $\hbar\omega = 0.6$ meV, some intensity is detected around $(0, 1, 0)$ above the background. However, this could stem from the tail of the high order contamination. At $\hbar\omega = 2$ meV two clear peaks appear.

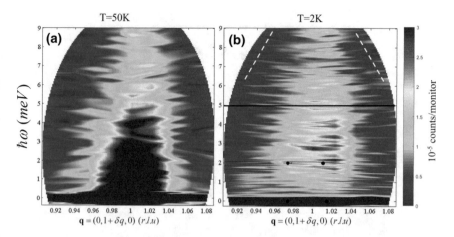

Fig. 4.1 SDW dispersion. False color map of normalized intensity as a function of neutron energy transfer $\hbar\omega$ and momentum transfer $\mathbf{q}$ at $T = 50$ K (**a**) and $T = 2$ K (**b**). The raw data is interpolated. The black horizontal line in panel (**b**) demonstrates a constant energy cut along which the intensity is integrated and plotted in Fig. 4.3. Dashed white lines in panel (**b**) represent cuts along which the background is determined. The black symbols indicate the center of the $\hbar\omega = 0$ and 2 meV peaks demonstrated in Fig. 4.2, and make the bottom part of an hourglass. At the right peak of $\hbar\omega = 2$ meV we had 50 counts per 2 million at the monitor

For fitting, the instrument was modeled using Popovici ResCal5 [18], and the resolution was calculated. Black horizontal lines in Fig. 4.2 represent the q-resolution at each energy. This was taken into account as a constant width Gaussian at each energy, which was convoluted with a Lorentzian (Voigt function). The fit with two Voigt functions is demonstrated in Fig. 4.2 by solid lines. The fit to the $\hbar\omega = 2$ meV data indicates a peak separation of 0.04 r.l.u. The same separation is found in the elastic peaks [2], as demonstrated in the inset. The peaks centers are illustrated in Fig. 4.1b by the solid points. The static and dynamic SDW correlation lengths, determined from the peaks width, are 85 ± 12 Å and 44 ± 5 Å respectively. With increasing energy to 4 meV and then to 6 meV, the two peaks are no longer resolved. However, the measured peak is asymmetric because of the two underlying incommensurate peaks coming closer together. At 8 meV the intensity diminishes. This behavior reminds two "legs" dispersing downwards from some crossing energy as in the hourglass.

To further investigate the inelastic behavior, we sum the intensity over $\mathbf{q}$ at constant energy cuts. The horizontal line in Fig. 4.1b presents one such cut. Background contribution is estimated from the data along the dashed diagonal lines in Fig. 4.1b, and subtracted. Figure 4.3 presents the background subtracted $\mathbf{q}$-integrated intensity versus energy transfer $\langle I \rangle (\omega) = \sum_{\mathbf{q}} I(\mathbf{q}, \omega)$, starting from $\hbar\omega = 0.15$ meV to avoid the high intensity elastic peak. At $T = 50$ K, $\langle I \rangle (\omega)$ monotonically grows as the frequency decreases. In contrast, at $T = 2$ K, $\langle I \rangle (\omega)$ reaches a maximum at some $\hbar\omega_{max}$ between 2 and 3 meV, and drops towards $\hbar\omega = 0$, although residual elastic

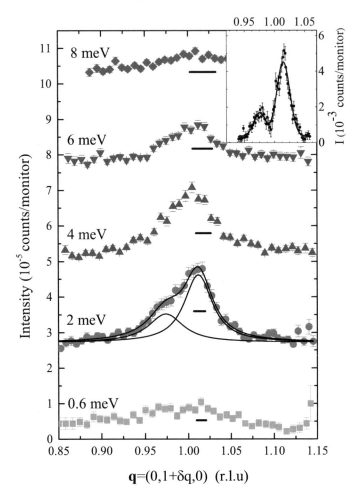

Fig. 4.2 Evolution of the SDW peaks with energy at $T = 2$ K. Momentum scan along k centered at $(0, 1, 0)$ for different energy transfers at $T = 2$ K. Scans are shifted consecutively by 2.5×10^{-5} counts/monitor for clarity. Inset: SDW elastic peaks for the same **q** scan also at $T = 2$ K. Background from higher temperature was subtracted. For energies of $\hbar\omega = 0$ and 2 meV, a sum of two Voigt functions is fitted to the data (solid black lines). The peak separation for $\hbar\omega = 2$ meV is 0.04 r.l.u, as in the $\hbar\omega = 0$ case (see inset). Black horizontal lines represents the instrumental resolution

scattering intensity is observed near $\hbar\omega = 0$. Measurements on $\mathrm{La}_{2-x}\mathrm{Ba}_x\mathrm{CuO}_4$ with $0.0125 \leq x \leq 0.035$ which were limited to energies below 1meV agree with our results [19]. This plot demonstrates more clearly the aforementioned soft gap in spin excitations which develops at low temperatures.

We summarize the available data on LSCO $x = 1.92\%$ in Fig. 4.4a. In this figure we show the temperature dependence of the **q**-integrated scattering intensity at three different energies. The data at $\hbar\omega = 0$ is taken from Ref. [2] and multiplied by

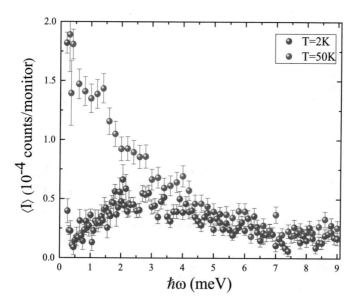

Fig. 4.3 q-integrated intensity versus neutron energy transfer at low (2 K) and high (50 K) temperatures. Integrated intensity is calculated for each energy as sum of the counts over **q** along horizontal
lines like the one shown in Fig. 4.1. Background is estimated from the counts along the two dashed
lines shown in Fig. 4.1a and subtracted from the raw data

2×10^{-3} for clarity. It shows that a long range static SDW appears at a temperature
of 30 K. The intensity at $\hbar\omega = 0.6$ meV increases as the temperature is lowered, peaks
at 38 K, and then decreases. This result demonstrates that dynamically fluctuating
SDW at $\hbar\omega > 0$ diminishes upon cooling before long range static incommensurate
order develops. The same effect, although less sharp, is observed for $\hbar\omega = 2$ meV
at 45 K.

Figure 4.4b depicts the temperature dependence of the nodal gap from Ref. [2] as
measured by ARPES. This gap opens at $T_{NG} = 45$ K, which is the same temperature
where the spectral density at $\hbar\omega_{max}$ begins to diminish. The maximum electronic gap
value Δ agrees with isolated dopant-hole bound state calculations [20]. We note that
$\hbar\omega_{max}$ and $k_B T_{NG}$ are of the same order of magnitude. Our result indicates a strong
link between the dynamically fluctuating SDW and the nodal gap.

In order to investigate whether CDW plays a role in the nodal gap [21], we con-
ducted a search for CDW in this sample by two different methods: off resonance X-ray
diffraction (XRD) and resonance elastic X-ray scattering (REXS). The experiments
were done at PETRA III on the P09 beam-line and at BESSY on the UE46-PGM1
beam-line, respectively. In REXS, the background subtraction is not trivial, so we
only present here our XRD data. Nonetheless, the final conclusion from both methods
is the same.

In Fig. 4.5 we show results from LSCO samples with $x = 1.92\%$, $x = 6.0\%$,
and $La_{2-x}Ba_xCuO_4$ (LBCO) $x = 12.5\%$. The data sets are shifted vertically for

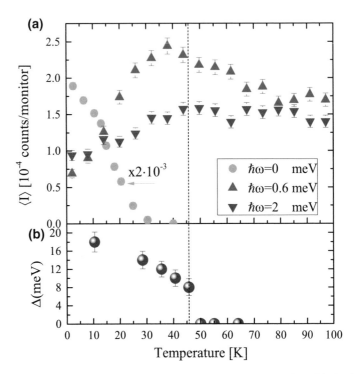

Fig. 4.4 Temperature dependence of all experimental parameters. **a** Elastic and inelastic incommensurate SDW intensities at different energies from neutron scattering. **b** ARPES measurement of the nodal gap at k_F [2]. The dashed vertical line emphasizes the fact that the nodal gap opens when the amplitude of dynamic spin fluctuations at $\hbar\omega \approx 2$ meV decreases

clarity. The LBCO sample is used as a test case, since it has well established CDW and presents strong diffraction peaks. The measurements were taken at 7 and at 70 K, which are below and above the CDW critical temperature of LBCO [22]. We performed two types of scans: a "stripes" scan along $(0, \delta q, 8.5)$ direction and a "checkerboard" scan along $(\delta q, \delta q, 8.5)$ direction. We chose to work at $l = 8.5$ to minimize contribution from a Bragg peak at $l = 8$ or $l = 9$. For LBCO at $T = 7$ K, there is a clear CDW peak at $\delta q = \pm 0.24$ in the "checkerboard" scan, which is absent at high temperatures. In contrast, for the LSCO samples there is no difference between the signal at high and low temperatures. Since δq of the CDW peak depends on doping, in our sample it is expected to be close to $\delta q = 0$, where a tail of the Bragg peak could potentially obscure the CDW peak. Arrows in Fig. 4.5 show where we might expect the CDW peaks, should they appear, based on linear scaling with doping. These positions are out of the $\delta q = 0$ peak tail, and not obscured. Thus, although we are in experimental conditions appropriate to find a CDW, it is not observed within our sensitivity. In fact, CDW is even absent at higher doping as demonstrated by our experiment with LSCO $x = 6\%$ sample. We observed the same null-result with

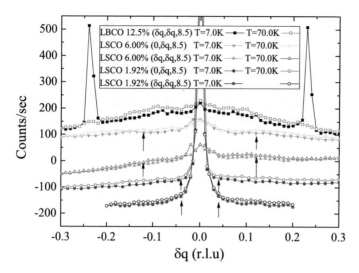

Fig. 4.5 Hard X-ray diffraction on three different samples: LSCO with $x = 1.92\%$ and $x = 6\%$, and LBCO $x = 12.5\%$. Scans are done in two different orientations and two different temperatures. CDW is detected only in LBCO

the REXS experiment. It is important to mention that hourglass excitations with no stripe-like CDW were observed previously [16].

Our main results are as follows: we find the bottom part of an hourglass dispersion inside the AFM phase of LSCO. The hourglass does not start from zero energy, but has a soft gap from the static SDW order. A CDW order seems to be absent in our sample. Upon cooling the system, a nodal gap in electronic excitations opens just when the strongest spin excitations start to diminish. It is therefore sufficient for the SDW fluctuations to slow down without completely freezing out in order to modify the band structure.

4.3 Methods

For the Neutron scattering experiment, the sample was mounted on aluminum holder covered with Cd foils, and oriented in the (h, k, 0) scattering plane. A Be filter was used to minimize contamination from high order monochromator Bragg reflections. The scattered neutrons are recorded with a nine bladed graphite analyzer. All the blades are set to scatter neutrons at the same final energy of 5 meV, and direct the scattered neutrons through an adjustable radial collimator to different predefined areas on a position sensitive detector [23, 24]. This monochromatic q dispersive mode allows for an efficient mapping of magnetic excitations with an excellent q resolution.

Two types of scans were used: (I) energy scan, in which the incoming neutrons energy is swept, and the **q** information is embedded in the position of each blade. (II) momentum scan, in which the incoming neutrons energy is fixed, the nine blades cover a small window in **q**, and the entire window is scanned. The contribution to a given **q** is a weighted sum from the different blades.

Despite the Be filter, some contribution from the nuclear structure is unavoidable. For elastic scattering, this contribution survives to higher temperature than does the magnetic part, and therefore can be easily subtracted. For inelastic scattering, the contribution from phonons could not be subtracted, but it is expected to vary slowly with temperature close to the magnetic phase transitions. Therefore, all features in this scattering experiment which show abrupt temperature dependence around and below $T = 50$ K are associated with the electronic (magnetic) system.

References

1. Kapon I, Ellis DS, Drachuck G, Bazalitski G, Weschke E, Schierle E, Strempfer J, Niedermayer C, Keren A (2017) Opening a nodal gap by fluctuating spin-density wave in lightly doped $La_{2-x}Sr_xCuO_4$. Phys Rev B 95:104512. https://doi.org/10.1103/PhysRevB.95.104512
2. Gil D, Elia R, Galina B, Amit K, Christof N, Ming S, Amit K (2014) Comprehensive study of the spin-charge interplay in antiferromagnetic $La_{2-x}Sr_xCuO_4$. Nat Commun 5:3390. https://doi.org/10.1038/ncomms4390
3. Harter JW, Maritato L, Shai DE, Monkman EJ, Nie Y, Schlom DG, Shen KM (2012) Nodeless superconducting phase arising from a strong (π, π) antiferromagnetic phase in the infinite-layer electron-doped $Sr_{1-x}La_xCuO_2$ compound. Phys Rev Lett 109:267001. https://doi.org/10.1103/PhysRevLett.109.267001
4. Vishik IM, Hashimoto M, He R-H, Lee W-S, Schmitt F, Lu D, Moore RG, Zhang C, Meevasana W, Sasagawa T et al (2012) Phase competition in trisected superconducting dome. Proc Natl Acad Sci 109:18332–18337. https://doi.org/10.1073/pnas.1209471109
5. Razzoli E, Drachuck G, Keren A, Radovic M, Plumb NC, Chang J, Huang Y-B, Ding H, Mesot J, Shi M (2013) Evolution from a nodeless gap to $d_{x^2-y^2}$-wave in underdoped $La_{2-x}Sr_xCuO_4$. Phys Rev Lett 110:047004. https://doi.org/10.1103/PhysRevLett.110.047004
6. Peng Y, Meng J, Mou D, He J, Zhao L, Wu Y, Liu G, Dong X, He S, Zhang J et al (2013) Disappearance of nodal gap across the insulator-superconductor transition in a copper-oxide superconductor. Nat Commun 4:2459. https://doi.org/10.1038/ncomms3459
7. Croft TP, Lester C, Senn MS, Bombardi A, Hayden SM (2014) Charge density wave fluctuations in $La_{2-x}Sr_xCuO_4$ and their competition with superconductivity. Phys Rev B 89:224513. http://link.aps.org/doi/10.1103/PhysRevB.89.224513
8. Matsuda M, Fujita M, Yamada K, Birgeneau RJ, Endoh Y, Shirane G (2002) Electronic phase separation in lightly doped $La_{2-x}Sr_xCuO_4$. Phys Rev B 65:134515. https://doi.org/10.1103/PhysRevB.65.134515
9. Niedermayer Ch, Bernhard C, Blasius T, Golnik A, Moodenbaugh A, Budnick JI (1998) Common phase diagram for antiferromagnetism in $La_{2-x}Sr_xCuO_4$ and $Y_{1-x}Ca_xBa_2Cu_3O_6$ as seen by muon spin rotation. Phys Rev Lett 80:3843–3846. https://doi.org/10.1103/PhysRevLett.80.3843
10. Squires GL (2012) Introduction to the theory of thermal neutron scattering. Cambridge University Press
11. Capati M, Caprara S, Di Castro C, Grilli M, Seibold G, Lorenzana J (2015) Electronic polymers and soft-matter-like broken symmetries in underdoped cuprates. Nat Commun 6:7691. https://doi.org/10.1038/ncomms8691

12. Hücker M, Christensen NB, Holmes AT, Blackburn E, Forgan EM, Liang R, Bonn DA, Hardy WN, Gutowski O, Zimmermann MV, Hayden SM, Chang J (2014) Competing charge, spin, and superconducting orders in underdoped $YBa_2Cu_3O_y$. Phys Rev B 90:054514. https://doi.org/10.1103/PhysRevB.90.054514

13. Tranquada JM, Woo H, Perring TG, Goka H, Gu GD, Xu G, Fujita M, Yamada K (2004) Quantum magnetic excitations from stripes in copper oxide superconductors. Nature 429:534–538. https://doi.org/10.1038/nature02574

14. Matsuda M, Fujita M, Wakimoto S, Fernandez-Baca JA, Tranquada JM, Yamada K (2008) Magnetic dispersion of the diagonal incommensurate phase in lightly doped $La_{2-x}Sr_xCuO_4$. Phys Rev Lett 101:197001. https://doi.org/10.1103/PhysRevLett.101.197001

15. Rømer AT, Chang J, Christensen NB, Andersen BM, Lefmann K, Mähler L, Gavilano J, Gilardi R, Niedermayer Ch, Rønnow HM, Schneidewind A, Link P, Oda M, Ido M, Momono N, Mesot J (2013) Glassy low-energy spin fluctuations and anisotropy gap in $La_{1.88}Sr_{0.12}CuO_4$. Phys Rev B 87:144513. https://doi.org/10.1103/PhysRevB.87.144513

16. Drees Y, Lamago D, Piovano A, Komarek AC (2013) Hour-glass magnetic spectrum in a stripeless insulating transition metal oxide. Nat Commun 4:2449. https://doi.org/10.1038/ncomms3449

17. Keimer B, Birgeneau RJ, Cassanho A, Endoh Y, Greven M, Kastner MA, Shirane G (1993) Soft phonon behavior and magnetism at the low temperature structural phase transition of $La_{1.65}Nd_{0.35}CuO_4$. Zeitschrift für Physik B Condensed Matter 91:373–382. https://doi.org/10.1007/BF01344067

18. Popovici M (1975) On the resolution of slow-neutron spectrometers. IV. The triple-axis spectrometer resolution function, spatial effects included. Acta Crystallogr Sect A Cryst Phys Diffr Theor Gen Crystallogr 31:507–513. https://doi.org/10.1107/S0567739475001088

19. Wagman JJ, Van Gastel G, Ross KA, Yamani Z, Zhao Y, Qiu Y, Copley JRD, Kallin AB, Mazurek E, Carlo JP, Dabkowska HA, Gaulin BD (2013) Two-dimensional incommensurate and three-dimensional commensurate magnetic order and fluctuations in $La_{2-x}Ba_xCuO_4$. Phys Rev B 88:014412. https://doi.org/10.1103/PhysRevB.88.014412

20. Sushkov OP, Kotov VN (2005) Theory of incommensurate magnetic correlations across the insulator-superconductor transition of underdoped $La_{2-x}Sr_xCuO_4$. Phys Rev Lett 94:097005. https://doi.org/10.1103/PhysRevLett.94.097005

21. Berg E, Chen C-C, Kivelson SA (2008) Stability of nodal quasiparticles in superconductors with coexisting orders. Phys Rev Lett 100:027003. https://doi.org/10.1103/PhysRevLett.100.027003

22. Tranquada JM, Gu GD, Hücker M, Jie Q, Kang H-J, Klingeler R, Li Q, Tristan N, Wen JS, Xu GY, Xu ZJ, Zhou J, Zimmermann MV (2008) Evidence for unusual superconducting correlations coexisting with stripe order in $La_{1.875}Ba_{0.125}CuO_4$. Phys Rev B 78:174529. https://doi.org/10.1103/PhysRevB.78.174529

23. Bahl CRH, Lefmann K, Abrahamsen AB, Rønnow HM, Saxild F, Jensen TBS, Udby L, Andersen NH, Christensen NB, Jakobsen HS et al (2006) Inelastic neutron scattering experiments with the monochromatic imaging mode of the RITA-II spectrometer. Nucl Instrum Methods Phys Res Sect B Beam Interact Mater Atoms 246:452–462. https://doi.org/10.1016/j.nimb.2006.01.023

24. Lefmann K, Niedermayer Ch, Abrahamsen AB, Bahl CRH, Christensen NB, Jacobsen HS, Larsen TL, Häfliger P, Filges U, Rønnow HM (2006) Realizing the full potential of a RITA spectrometer. Phys B Condens Matter 385:1083–1085. https://doi.org/10.1016/j.physb.2006.05.372

Chapter 5
Conclusions

We demonstrated that the Stiffnessometer can measure penetration depth two orders of magnitude longer, or stiffness four orders of magnitude smaller, than ever before. This allows us to perform measurements closer to T_c and explore the nature of the superconducting phase transition, or determine the stiffness at low T in cases where it is naturally very weak. The Stiffnessometer also allows measurements of very small critical current or long coherence lengths. The measurements are done in zero magnetic field with no leads, thus avoiding demagnetization, vortices, and out-of-equilibrium issues.

Using this new magnetic-field-free superconducting stiffness technique, we shed new light on the SC phase transition in LSCO $x = 1/8$. In this compound, there is a temperature interval of 0.7 K where SC current can flow in the CuO_2 planes but not between them. That is, in this temperature range, the in-plane stiffness if finite while the interplane stiffness is zero. When stiffness develops in both directions, the ratio of penetration depths obeys $\lambda_c/\lambda_{ab} \geq 10$.

Thus, studying superconducting properties with the stiffnessometer leads to new results. The most intruiging one is two stiffness transition temperatures in the same sample.

Moreover, studying the relationship between spin and charge degrees of freedom in extrem underdoped LSCO, we found the bottom part of an hourglass dispersion inside the AF phase. The hourglass does not start from zero energy, but has a soft gap from the static SDW order. A CDW order seems to be absent in our sample. Upon cooling the system, a nodal gap in electronic excitations opens just when the strongest spin excitations start to diminish. It is therefore sufficient for the SDW fluctuations to slow down without completely freezing out in order to modify the band structure.

© Springer Nature Switzerland AG 2019

I. Kapon, *Searching for 2D Superconductivity in La2−xSrxCuO4 Single Crystals*,
Springer Theses, https://doi.org/10.1007/978-3-030-23061-6_5

Appendix
Vector Potential of Finite Coil

The finite coil is treated as a collection of current loops. The radius of the loops and the distance between them is taken from the specs of the coil itself. The radius is taken to be at the center of the wire. For a circular loop of radius R lying in the xy plane centered around the origin and carrying a current I, the vector potential has only ϕ component (cylindrical coordinates), and is given by

$$A_\phi(\rho, z) = \frac{\mu_0}{4\pi} \frac{4Ia}{\sqrt{R^2 + \rho^2 + z^2 + 2R\rho}} \left[\frac{(2 - k^2)K(k^2) - 2E(k^2)}{k^2} \right]$$

where K, E are the complete elliptic integrals of the first and second kind, and

$$k^2 = \frac{4R\rho}{R^2 + \rho^2 + z^2 + 2R\rho}$$

(taken from [1] page 182).

Note that here for the elliptic integrals I use Matlab notation, where

$$K(m) = \int_0^{\pi/2} \frac{d\theta}{\sqrt{1 - m\sin^2\theta}} = \int_0^1 \frac{dt}{\sqrt{(1 - t^2)(1 - mt^2)}}$$

$$E(m) = \int_0^{\pi/2} d\theta\sqrt{1 - m\sin^2\theta} = \int_0^1 dt\sqrt{\frac{1 - mt^2}{1 - t^2}},$$

whereas Jackson, for argument of m will use m^2 in the integrand, i.e. $K_{Matlab}(m) = K_{Jackson}(\sqrt{m})$.

Thus, the vector potential of the coil of length L with wires of diameter d is given by

$$A_{coil}(\rho, z) = \sum_R \sum_{j=1}^{L/2d} A_{loop,R}(\rho, z - jd) + A_{loop,R}(\rho, z + jd)$$

© Springer Nature Switzerland AG 2019
I. Kapon, *Searching for 2D Superconductivity in La$_{2-x}$Sr$_x$CuO$_4$ Single Crystals*,
Springer Theses, https://doi.org/10.1007/978-3-030-23061-6

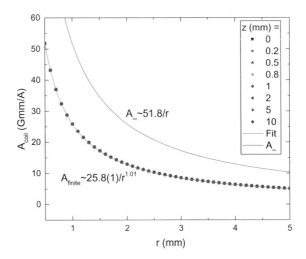

Fig. A.1 Finite versus infinite coil vector potential

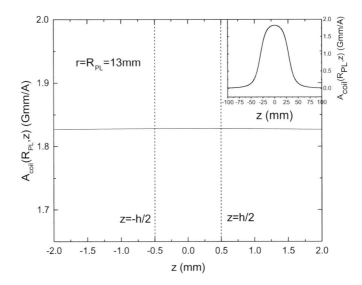

Fig. A.2 Finite coil vector potential z-dependence at the pick-up loop position

Figure A.1 presents the vector potential of a finite coil as a function of the distance from its center, r, in mm, for different z positions. The coil is 60 mm high, has 0.54 mm polyamide core and two layers of windings with a 0.05 mm diameter wire. The potential drops off as $1/r$ like in the case of infinite coil, which appears in the figure in blue line. The different prefactor does not affect the Stiffnessometer PDE and its solution, as everything is normalized by $A_{IC}(R_{PL})$, and the absolute value of the vector potential is measured by ΔV_{IC}^{max}.

The inset of Fig. A.2 shows the vector potential of the same coil at the pickup loop position $r = R_{PL} = 13$ mm as a function of the height z. The main figure is a zoom-in on the ring area from $z = -h/2$ to $z = h/2$. This demonstrates that the vector potential is z-independent for that region, as expected from infinite coil.

Reference

1. Jackson JD (1998) Classical electrodynamics, 3rd edn. Wiley

Printed in the United States
By Bookmasters